Explosive Remnants of War

Mitigating the Environmental Effects

sipri

Stockholm International Peace Research Institute

SIPRI is an independent institute for research into problems of peace and conflict, especially those of arms control and disarmament. It was established in 1966 to commemorate Sweden's 150 years of unbroken peace.

The Institute is financed by the Swedish Parliament. The staff, the Governing Board and the Scientific Council are international.

The Board and Scientific Council are not responsible for the views expressed in the publications of the Institute.

Governing Board

Rolf Björnerstedt, Chairman (Sweden)
Egon Bahr (FR Germany)
Francesco Calogero (Italy)
Tim Greve (Norway)
Max Jakobson (Finland)
Karlheinz Lohs (German Democratic Republic)
Emma Rothschild (United Kingdom)
The Director

Director

Frank Blackaby (United Kingdom)

Stockholm International Peace Research Institute
Pipers väg 28, S-171 73 Solna, Sweden
Cable: Peaceresearch Stockholm
Telephone: 08-55 97 00

Explosive Remnants of War
Mitigating the Environmental Effects

Edited by
Arthur H. Westing

sipri

Stockholm International Peace Research Institute

United Nations Environment Programme

Taylor & Francis
London and Philadelphia
1985

UK	Taylor & Francis Ltd, 4 John Street, London WC1N 2ET
USA	Taylor & Francis Inc., 242 Cherry Street, Philadelphia, PA 19106-1906

British Library Cataloguing in Publication Data

Explosive remnants of war: mitigating the
    environmental effects.
    1. Bomb reconnaissance
    I. Westing, Arthur H.   II. Stockholm International Peace
    Research Institute.   III. United Nations,
    Environmental Program
    358'.2   UF767
    ISBN 0-85066-303-2

Library of Congress Cataloging in Publication Data is available

Cover design by Malvern Lumsden
Typeset by Alresford Phototypesetting, Alresford, Hants.
Printed in Great Britain by Redwood Burn Ltd,
Trowbridge, Wilts.

Attribution

Explosive Remnants of War: Mitigating the Environmental Effects has been prepared by SIPRI as a project within the SIPRI/UNEP programme on 'Military
activities and the human environment'. The present volume is an outgrowth
of a meeting of a group of high-level experts convened by UNEP (with the
assistance of SIPRI) in Geneva, 25–28 July 1983.

The views expressed in this book are not necessarily those of either SIPRI
or UNEP.

This book represents the third in a series that has been produced under the
SIPRI/UNEP programme. The two prior titles are:

Environmental Warfare: A Technical, Legal and Policy Appraisal (edited by A.
H. Westing; Taylor & Francis, London, 108 pp.; 1984); and

Herbicides in War: The Long-term Ecological and Human Consequences (edited
by A. H. Westing; Taylor & Francis, London, 210 pp.; 1984).

Preface

This book is the third major product of a research programme into the environmental consequences of military activity, jointly financed by SIPRI and the United Nations Environment Programme (UNEP). The programme, which extends over a number of years, is based at SIPRI under the leadership of Dr Arthur H. Westing (Adjunct Professor of Ecology at Hampshire College in Masssachusetts, USA), who is an international authority on these matters.

The book will, it is hoped, draw attention to the explosive munitions that remain hidden in former theatres of war and which claim so many innocent victims—often children—for so many years. The authors of the book recommend a number of technical, legal and policy approaches to mitigating this pernicious environmental problem. Among the technical approaches is the widespread adoption of munitions that become de-activated automatically following their intended hostile purpose. Among the legal approaches is the universal adherence to the Inhumane Weapons Convention of 1981. And among the policy approaches is the identification or establishment of a relevant information and training centre, perhaps under UNEP auspices.

The explosive remnants of war—a horrifying component of the human environment—have long been a major concern of UNEP. The SIPRI/UNEP collaboration thus represents another step towards the alleviation of this pernicious problem.

<table>
<tr><td>Mostafa K. Tolba</td><td>Frank T. Blackaby</td></tr>
<tr><td>Executive Director</td><td>Director</td></tr>
<tr><td>UNEP</td><td>SIPRI</td></tr>
<tr><td>April 1985</td><td></td></tr>
</table>

Acknowledgements

The editor is pleased to acknowledge the very able research assistance of Carol Stoltenberg-Hansen, editorial assistance of Jetta Gilligan Borg and secretarial assistance of Sandra Owen Bendtz.

Contents

Glossary and units of measure

I. Glossary

Cyclonite: Hexahydro-1,3,5-trinitro-s-triazine = RDX.

Influence mine: A mine set off by such influences as the target's magnetic field, by the noise the target generates (its acoustic emanation), by the change in water pressure the target produces, or by other remote signal from the target.

SIPRI: Stockholm International Peace Research Institute.

TNT: 2,4,6-trinitrotoluene.

UN: United Nations.

UNEP: United Nations Environment Programme (Nairobi).

UNGA: United Nations General Assembly (New York).

UNITAR: United Nations Institute for Training and Research (New York).

II. Units of measure

The units of measure and prefixes (and the abbreviations) employed in the text are in accordance with the international system (SI) of units (Goldman & Bell, 1981).

are(a) = 100 square metres = 1 076.39 square feet.

centi- (c-) = $10^{-1} \times$.

centimetre (cm) = 0.1 metre = 0.393 701 inch.

gram (g) = 10^{-3} kilogram = $2.204\,62 \times 10^{-3}$ pound.

hectare (ha) = 10^4 square metres = 0.01 square kilometre = 2.471 05 acres.

hect(o)- (h-) = $100 \times$.

hour (h) = 3 600 seconds.

joule (J) = 0.238 846 calorie.

kilo- (k-) = $10^3 \times$.

kilogram (kg) = 2.204 62 pounds.

kilometre (km) = 10^3 metres = 0.621 371 statute mile = 0.539 957 nautical mile.

kilometre, square (km²)=10⁶ square metres=100 hectares=247.105 acres=0.386 102 square statute mile=0.291 553 square nautical mile.

mega- (M-)=10⁶ ×.

megajoule (MJ)=10⁶ joules=238 846 calories.

metre (m)=3.280 84 feet.

metre, cubic (m³)=10³ litres=264.172 US gallons=219.969 British gallons=1.307 95 cubic yards.

metre, square (m²)=10.763 9 square feet=1.195 99 square yards.

milli- (m-)=10⁻³ ×.

milligram (mg)=10⁻⁶ kilogram=35.274 0 × 10⁻⁶ ounce.

millimetre (mm)=10⁻³ metre=0.039 370 1 inch.

second (s): see Goldman & Bell (1981, p. 3).

tonne (t)=10³ kilograms=1.102 31 US (short) tons=0.984 207 British (long) ton.

Reference

Goldman, D. T. & Bell, R. J. (eds). 1981. *International system of units (SI)*. Washington: US National Bureau of Standards Special Publ. No. 330, 48 pp.

Introduction

Many of the fatalities and maimings during the course of a war result from high-explosive munitions. It is an inevitable though not widely appreciated fact that a significant fraction of these high-explosive munitions do not explode during the war, but remain dangerously ready to do so after the war is over. This terrible post-war legacy continues to result in fatalities and maimings for many years or decades following the cessation of hostilities.

The present multi-authored volume begins with an overview of the explosive-remnant problem (chapter 1) and then provides a number of actual case studies: the aftermath of World War II in Poland (chapter 2) and in Libya (chapter 3); and the aftermath of the Second Indochina War in Viet Nam and Laos (chapter 4). The book continues with the extent to which the problem can be mitigated by technical means, both in the terrestrial environment (chapter 5) and the marine environment (chapter 6). The use of dogs in detecting explosive remnants is singled out for special attention (chapter 7). The book concludes with a legal analysis of the problem (chapter 8).

The text of the volume is complemented by a selection of technical, legal and policy references for further reading (appendix 1). Next comes a chronology of relevant United Nations activities (appendix 2). Copies of the treaties most relevant to the subject are also provided for the convenience of the reader (appendices 3–7).

The book is an outgrowth of a select symposium convened in Geneva on 25–28 July 1983 by the United Nations Environment Programme with the assistance of SIPRI. That symposium led to a report which is reproduced here (appendix 8). A number of the participants of that symposium have elaborated upon their informal symposium contributions for inclusion in the present volume, together with the works of several others who were not involved in the symposium. The authors of this book are:

Colonel *Bengt Anderberg* (Armoured-troop School, Swedish Army, 541 29 Skövde, Sweden), an authority on explosive remnants of war and on international humanitarian law.

Mr *Jozef Goldblat* (Stockholm International Peace Research Institute, 171 73 Solna, Sweden), an authority on arms control law and policy.

Murray Hiebert (Indochina Project, Washington, DC 20002, USA), an authority on rural Laos.

Professor *Robert E. Lubow* (Department of Psychology, University of Tel Aviv, Tel Aviv 69978, Israel), an authority on dog behaviour and training.

Earl S. Martin (Mennonite Central Committee, Akron, PA 17501, USA), an authority on rural Viet Nam.

Professor *Boguslaw A. Molski* (Botanical Garden, Polish Academy of Sciences, 02 973 Warsaw, Poland, an authority on environmental protection.

Commander *Dewitt H. Moody* (US Navy [Retd]; Searle Consortium, Alexandria, VA 22302, USA), an authority on explosives and explosive-ordnance disposal.

Colonel *Jan Pajak* (Combat Engineer Headquarters, Polish Army, Warsaw, Poland), an authority on explosive remnants of war.

Captain *Willard F. Searle, Jr* (US Navy [Retd]; Searle Consortium, Alexandria, VA 22302, USA), an authority on ocean engineering and salvage.

Professor *Khairi Sgaier* (Department of Agronomy, University of Alfateh, Tripoli, Libya), an authority on agriculture in arid areas.

Professor *Arthur H. Westing* (Stockholm International Peace Research Institute, 171 73 Solna, Sweden), an authority on environmental impact of military activities.

1. Explosive remnants of war: an overview

Arthur H. Westing
Stockholm International Peace Research Institute

I. Introduction

The tragic sequelae of war are likely to include such material remnants as unexploded land and sea mines, and they invariably include a diversity of other unexploded munitions such as bombs, shells, rockets and grenades. Often hidden from view and randomly distributed in huge numbers, this plethora of life-threatening devices poses severe problems in post-war rehabilitation and utilization of the land and often results in calamitous accidents.

This chapter provides an overview of the environmental impact of the explosive remnants of war, that is, of the post-war residuum of unexploded mines and other munitions (duds). This is a huge and growing problem owing to the vast and ever increasing quantities of explosive ordnance that are expended in modern warfare, often in an indiscriminate fashion, and of which of the order of 10 per cent fail to explode as planned; and because of the immense numbers of land or sea mines, designed to withstand environmental deterioration, that are emplaced or scattered in a fashion meant to defy deliberate detection. The chapter provides the military background (regarding both land and sea) necessary to understand the use of the pertinent ordnance. It illustrates the dimensions of the problem through brief apropos descriptions of World War II, the Second Indochina War, the Arab–Israeli Wars of 1967 and 1973, the Israeli incursion into Lebanon of 1982, and the Falklands/Malvinas War of 1982. It goes on to outline the existing means of disposal, both on land and at sea, stressing the dangers and complexities involved. The chapter concludes with some technical, legal and other recommendations to mitigate the problem, among them suggestions that all explosive ordnance be designed so as to become automatically harmless in due course; that an open clearing house, information repository and research and training centre be established, presumably under United Nations auspices; that co-operative multinational clean-up programmes be initiated;

and that the Inhumane Weapon Convention of 1981 (see appendix 3) be widely adopted. A select bibliography of relevant background information is provided elsewhere (see appendix 1).

II. Military background

Potentially explosive remnants of war have a number of origins. To begin with, there are the bombs, artillery and mortar shells, rockets and grenades that malfunctioned at the time they were expended, the so-called duds. Then there are the sea mines, land mines and booby traps that were emplaced, but not subsequently triggered or removed during the war. Other miscellaneous sources include abandoned ammunition dumps and caches, dumpings of unwanted munitions (often at sea or in lakes) and abandoned vehicles, sunken ships or downed aircraft containing explosive devices.

Duds

Modern warfare—whether land, sea or air—depends upon the expenditure of vast and seemingly ever growing numbers of high-explosive munitions. Indeed, there are indications that the fire-power of our military forces has been growing at an essentially exponential rate during the twentieth century (Lumsden, 1978, p. 45; Westing, 1980, p. 3). By way of recent example, it has been calculated that during the Second Indochina War of 1961–1975, the USA dropped approximately 20 million bombs of various sizes and fired some 230 million artillery (including naval) shells; additionally expended were countless millions of rockets, mortar shells and grenades (Westing, 1980, p. 96). A recent ordnance innovation that helps to inflate these numbers was the so-called cluster-bomb unit, a large cannister dropped like a bomb which then pops open to release and scatter scores of individual high-explosive, grenade-sized or larger bomblets (Krepon, 1973–1974). Technical descriptions of many of the high-explosive munitions in question are available, for example, for bombs (Hyman, 1982; Marriott, 1975–1976; Pretty, 1984–1985, pp. 382–423), artillery shells (Foss, 1984–1985, pp. 692–737), mortar shells (Goad & Halsey, 1982, pp. 169–177; Hogg, 1984–1985, pp. 565–656; Owen, 1979, pp. 215–251), rockets (Hogg, 1984–1985, pp. 686–748; Hyman, 1982; Owen, 1979, pp. 157–214), grenades (Goad & Halsey, 1982, pp. 193–201; Hogg, 1984–1985, pp. 414–475; Owen, 1979, pp. 252–291), and various anti-personnel devices (British Ministry of Defence, 1980; Goad & Halsey, 1982; Lumsden, 1978; Prokosch, 1976).

Whereas the vast majority of the high-explosive munitions expended in warfare bursts at the time of delivery and thus poses no continuing danger, a small but significant proportion does not. The limited authoritative information available on this subject follows. During World War II from 5 per cent to 10 per cent of all US bombs failed to explode, those with delayed-action fuses accounting for the majority of these duds (War, 1943, p. 1). During the Second Indochina War US artillery shells equipped with the standard point-detonating fuse failed to explode 2.5 per cent of the time when set in the super-quick mode and from 5 per cent to 50 per cent of the time when set in the delay mode (Mahon, 1972, p. 198). US mortar shells did not detonate 10 per cent to 20 per cent of the time during the dry season, and 30 per cent or more of the time during the wet season (Swearington, 1969, p. 5). US hand grenades were duds 15 per cent to 25 per cent of the time during the dry season and 40 per cent to 50 per cent of the time during the wet season (Swearington, 1969, p. 3). The overall failure rate of all high-explosive munitions expended by the USA during the Second Indochina War was estimated to be of the order of 10 per cent (Swearington, 1969, p. 7).

If one employs a dud-rate value of 10 per cent for purposes of rough estimation in the case of Indochina, one can see that of the order of 2 million bombs, 23 million artillery shells, and many tens of millions of other high-explosive munitions did not explode as intended. An unknown fraction of these were salvaged for re-use during the war, often being remanufactured into mines or booby traps (Lowe, 1968; Milling, 1969; Vehnekamp, 1970); a further unknown fraction was, or has become, sufficiently defective so that the devices will never blow up; and a final unknown fraction remains clearly visible and can thus be avoided and destroyed with relative ease. On the other hand, many millions of unexploded munitions remain hidden for long periods as potentially lethal or maiming remnants of that war.

Mines

In addition to the high-explosive remnants mentioned above, modern land and sea forces generally depend upon the use of high-explosive mines. These are usually emplaced and often constructed so as to defy premeditated discovery, but are of course designed so as to detonate when inadvertently disturbed. Huge numbers of these devices are routinely employed in land warfare (Alder, 1980; Foss, 1979; Halloran, 1972; Kitching, 1975; Rybicki, 1984; Stampfer, 1981; Watson, 1961) as well as in naval warfare (Berry, 1979; Hartmann, 1979; Horne, 1982; Lancesseur, 1985; Marriott, 1974–1975; Niemann, 1983; Patterson, 1970–1971; Rouarch, 1984; Taylor, 1977; Whelan, 1980; Wile, 1982). Mines can be emplaced primarily for defensive purposes, that is, to prevent an attack or border crossing or to deny an area to the enemy, or they can be employed primarily for offensive (or harassment) purposes, that is, to bring about (directly or indirectly) enemy casualties or the destruction of matériel. In practice, however, it may be difficult to distinguish between these two strategies. In any case, many (if not most) of the mines emplaced during a war remain as a post-war remnant to plague the recipient nation long after the cessation of hostilities.

Land mines

Land mines fall into one of two general categories: anti-tank (or anti-vehicle) devices, which are generally blast weapons containing an explosive charge of perhaps 5–10 kilograms which is set off by pressure or such other means as a change in the magnetic field; and anti-personnel devices, which are generally fragmentation weapons containing an explosive charge of somewhat less than 0.5 kilogram which is set off by pressure, by disturbing a trip wire, or in some cases by vibration or other means. Technical descriptions of land mines are available (Crèvecoeur, 1977; Foss & Gander, 1984, pp. 152–214; Goad & Halsey, 1982, pp. 205–220; Hogg, 1981; Owen, 1979, pp. 304–339; Red Cross, 1973, pp. 49–51; Tresckow, 1975). Some anti-personnel mines jump (bound) up a metre or so before exploding. Some very small anti-personnel mines exist that depend upon the blast effect of perhaps 30–50 grams of explosive; these are meant to be scattered in very large numbers and, when stepped upon, to amputate the foot. A booby trap is an anti-personnel mine that is not hidden, but rather is disguised as a harmless and attractive object; its primary purpose is to harass an enemy (Brendt, 1967; Wildrick, 1969). Anti-personnel mines are usually laid among anti-tank mines; and regular mines of any sort can also be booby-trapped in order to make their neutralization more difficult and hazardous.

The location of a minefield is often, but by no means always, recorded by the side that lays it for post-war publication; under some circumstances its location is even announced during the time of hostilities. However, a recent militarily attractive innovation in land mines and their delivery systems permits the remote delivery of either anti-personnel or anti-tank mines in large numbers (Andrews, 1978; Army, 1984, pp. 50–51, 54–55, 72–73; Chase, 1980; Howell, 1977; McDavitt, 1979). These scatterable mines can be disseminated from tank- or truck-mounted devices in which case their distribution is more or less well controllable and the minefield could still be demarcated to a certain level of satisfaction. However, they can also be delivered into enemy territory by means of artillery or aircraft in which case the ability to record their location often becomes haphazard if not impossible. It is fortunate that these mines are not buried and also that at least some of the scatterable mines now being produced have built-in mechanisms that destroy them after a set time.

In the various North African campaigns during World War II, Germany, Italy, the United Kingdom and France laid many millions of land mines, mostly anti-tank mines (Ceva, 1981; Libya, 1981; see also chapter 3)—some 5 million according to one authority (Cestac, 1981, p. 12) and fully 19 million according to another (Watson, 1961). In the North African campaigns during World War II 18 per cent of the tanks destroyed were knocked out by mines; in the Italian campaign 28 per cent; and in the Pacific theatre 34 per cent (Stampfer, 1981).

Sea mines

Sea mines are either dense enough to sink and thus rest on the ocean floor, or else they are made to be buoyant and are then moored to the seabed; drifting (untethered, floating) sea mines are apparently not in intentional use by most nations (Hartmann, 1979; Lancesseur, 1985; Marriott, 1974–1975; Pretty, 1984–1985, pp. 198–228; Wettern, 1979). Sea mines are blast weapons, with the moored mines containing an explosive charge of perhaps 250 kilograms and the seabed mines one of perhaps 750 kilograms (Wilcke, 1971). Various ones can be set off by contact (either direct or with trailing antennae), by noise (acoustic radiation), by changes in water pressure or by changes in the magnetic field; those not set off by contact are referred to as influence mines.

Two examples of sea mining are provided here in order to suggest the magnitude of such operations. During World War II the USA laid nearly 31 000 sea mines in the Pacific Ocean against Japan (Berry, 1979). This is said to have resulted in the destruction of almost 1 100 ships, or one ship for every 28 mines emplaced. Then during the Second Indochina War the USA mined Haiphong harbour from the air with 8 000 sea mines and the navigable inland waters of North Viet Nam with an additional 3 000 (New York Times, 1973). The mining of Haiphong harbour achieved its objective of paralysing that facility (Luckow, 1982).

III. The post-war legacy

The residuum of unexploded ordnance of all sorts that remains mortally dangerous following a war continues for decades to result in tragedies, often involving children. In Libya, for example, during the four decades since World War II it has continued to be no rare incident for shepherds to step on the old buried land mines and be killed in the explosion that follows (Cestac, 1981, p. 13; Libya, 1981; Oliver, 1945–1946, p. 33). The following three recent newspaper headlines—also all relating to World War II munitions—serve to further illustrate this point: "Old Bomb Kills 3 in Japan" (Associated Press, 1974); "Old Shell Kills 21 [in Burma]" (United Press International, 1976); and "5 French Children Die as Old Shell Explodes" (Associated Press, 1981). As a result, countries on whose soil a war has been fought must maintain highly-trained munition disposal units whose work continues unabated the year round for decades. Disposing of high-explosive remnants of war is a grim business even for these highly trained disposal units. For example, Egyptian disposal units in a recent operation experienced one fatality for every 7 000 land mines cleared, that is, 143 fatalities per million mines cleared (Graves, 1975, p. 808). It has been suggested that in general one individual will be killed and two injured for each 5 000 land mines removed (Cestac, 1981, p. 23), although the experience of Poland indicates that the fatality rate need not be quite that high (see chapter 2). Examples of the material remnant problem from a number of specific wars follow, beginning with a brief further elaboration of World War II.

World War II

In the years since the end of World War II, French disposal units have been employing at any one time a total of about 80 specialists who clear high-explosive munitions on a continuing basis; more than 13 000 were cleared in 1978 alone (Cestac, 1981, p. 17). In the Netherlands a team of 90 demolition experts continues to handle about 2 000 wartime ordnance disposal cases annually and cannot keep up with the demand for its services (Associated Press, 1984). So far a total of 75 members of the Dutch disposal unit have been killed in the line of duty, as well as 210 German prisoners of war who were forced into clearing service at the end of the war. Soviet disposal units have since the end of World War II destroyed many tens of millions of explosive remnants with the help of huge numbers of troops (Evangelista, 1982–1983, p. 130; Ryabchikov, 1977). Neutralized since the end of that war in West Berlin alone have been more than 7 000 bombs, more than 748 000 artillery shells, and almost 476 000 grenades and other small explosive devices (Associated Press, 1977a). Finnish disposal units have since the end of World War II thus far disposed of over 6 000 bombs, 805 000 artillery shells, 66 000 mines, and 370 000 miscellaneous high-explosive munitions (Finnish Defence Forces, 1976, app. 4). Since World War II ended Japanese disposal units have so far eliminated more than 6 000 sea mines from their coastal waters, this effort roughly estimated to now be only half completed (Reuters, 1975). Most recently, these mines sank a ship in 1972 and another in 1975. Among the nations still burdened with extraordinarily many explosive remnants of World War II must be included especially Poland (Anderberg, 1981; see also chapter 2). Since that war Poland has disposed of more than 88 million items of explosive ordnance (of which about 15 million have been mines). The task continues, with over 200 000 items of various different types still being located and destroyed each year.

Second Indochina War

As noted earlier, vast amounts of munitions were expended by the USA during the Second Indochina War (see also chapter 4). These resulted in uncounted millions of dangerous duds, among them a grotesque variety of delayed-action anti-personnel weapons (Krepon, 1973–1974; Lumsden, 1978; Prokosch, 1976). Both sides employed anti-tank mines, anti-personnel mines and booby traps so that huge numbers of these devices also remain. A few eye-witness accounts will serve to illustrate the human dimensions of the post-war problem. An observer from South Viet Nam wrote that "there is a serious problem from things like mine fields, unexploded bombs, unexploded artillery rounds all over the place. There are places in the highlands where the Montagnards will not even go into the forest to hunt because of bombs lying around. Lots of people come into provincial hospitals with wounds from stepping on old mines or who somehow detonated unexploded rounds. It is a problem all over the rural areas" (Hickey, 1973). In visits to Quang Tri other observers found that injuries from previously unexploded munitions constituted the most serious medical problem in the province (Luce, 1974) and that some 300 people and 1 000 water buffaloes had been thus killed during the prior 12 months (Fonda, 1974). Various similar accounts could be cited from Viet Nam (Associated Press, 1977b; Martin, 1973) and Laos (Hiebert & Hiebert, 1978a; 1978b). Indeed, there are large portions of Indochina where there seems to be no peasant family that cannot recount a personal tragedy—whether of death or maiming—caused by previously unexploded munitions (Westing, 1975; 1980, pp. 95–96).

As indicated earlier, in 1972 the USA mined Haiphong harbour in North Viet Nam with some 8 000 sea mines as well as various navigable inland waters of North Viet Nam with an additional 3 000 sea mines (Luckow, 1982; New York Times, 1973). In 1973 the USA had to sweep Haiphong harbour (see appendix 7), a five-month job by a large naval task force that was not as difficult as it might have been inasmuch as the mines had been laid by the USA with subsequent relocation in mind (McCauley, 1974). The mines sown in the inland waters did not have to be sought out because they had been set to destroy themselves or become inert after a time.

Arab–Israeli Wars of 1967 and 1973

The Suez Canal and its environs were fought over in 1967 and the canal was closed by sunken ships and mines. The Suez Canal zone was again the scene of battle in 1973. As a result of these events immense numbers of undetonated high-explosive munitions got into the canal and its terrestrial margins, and the marine approaches were also mined. It required a huge, sophisticated and dangerous series of aerial, surface and subsurface operations by Egypt, the USA, the United Kingdom, France and the USSR over a period of more than a year to finally render the canal and its approaches sufficiently safe to be dredged and re-opened (Boyd, 1976; Graves, 1975; Pengelley, 1974; Searle & Moody, 1981, pp. 19–24; Studenikin, 1975). The United Kingdom contingent during a single five-month period found resting on one stretch of the canal bed, and neutralized, 516 anti-personnel mines, 125 anti-tank mines, 16 bombs, 9 cluster bombs, 508 bomblets, 234 artillery shells, 141 anti-tank rockets, 190 grenades, and many hundreds of miscellaneous additional items of explosive ordnance (Pengelley, 1974). All told, some 8 500 divers items of explosive ordnance were found in the canal and disposed of (Searle & Moody, 1981, pp. 20–21). Moreover, the Egyptian contingent cleared nearly 700 000 land mines from the terrain adjacent to the canal (Graves, 1975).

Many of these mines were non-metallic and had to be located manually (with, as noted earlier, substantial loss of life) since the best available electronic detectors had not been adequate for the job (Pecori, 1981).

Israeli incursion into Lebanon of 1982

The aftermath of the recent Israeli incursion into Lebanon includes the location and disposal problems associated with large numbers of many different types of high-explosive remnants. A joint US–Lebanese munition disposal unit (one of several) in a six-week period unearthed 250 different kinds of explosive ordnance, including more than a dozen bombs, some 200 bomblets, and hundreds of mines and grenades (Associated Press, 1982c). Forty-five bomblets were disposed of in the yard of an orphanage after an explosion killed four children and wounded five others. As noted earlier, children are often the victim of such tragedies (Associated Press, 1979). Several US and French disposal personnel have also been killed so far in the line of duty.

Falklands/Malvinas War of 1982

The recent Falklands/Malvinas conflict has left a legacy of many thousands of undetonated high-explosive munitions as well as many additional thousands of mines and numerous booby traps (Associated Press, 1982a; 1982b; Feron, 1982; Hill, 1984; Kelly, 1983; McWhirter, 1982). Thousands of small plastic anti-personnel mines were scattered indiscriminately from helicopters in unrecorded locations. Disposal units have already cleared thousands of the explosive remnants of this brief war, but the work is expected to continue for at least another year in the farming regions and for years beyond that in the peat bogs (which must be exploited for fuel) and other rural regions. At least one of the disposal personnel has been killed so far and several have lost limbs (Associated Press, 1982a; 1982b; McWhirter, 1982; United Press International, 1983).

Military firing ranges

A problem related to that of the material remnants of war under discussion here, and thus worthy of passing mention, is the reclamation for civil pursuits of shelling and bombing target areas which had been used by military forces for training purposes. The clean-up of such areas can be an extraordinarily difficult task. One example is the small Hawaiian island of Kahoolawe (20°20′N 156°40′W) that was long used by the US Navy in this way and which the local inhabitants now want to have rehabilitated (Courson, 1972; LeBarron & Walker, 1971; Time, 1977).

IV. Means of disposal

Disposal on land

The neutralization of land mines and other unexploded munitions requires specialized training and nevertheless remains an agonizingly tedious and dangerous process (Golino, 1984; Lowe, 1968; Mercer, 1969; see also chapter 5). First these life-threatening remnants must be located and then they must be rendered safe. Mines are often specifically designed and usually specifically emplaced so as to make detection impossible; and dud munitions are often randomly concealed just below the surface.

Electronic metal detectors work reasonably well under some conditions, but are, of course, useless in locating non-metallic mines. Some non-metallic (or combination metallic/non-metallic) mine detectors exist, but are unreliable. Technical descriptions of electronic mine detectors are available (British Ministry of Defence, 1980, pp. 420–423; Foss & Gander, 1984, pp. 224–237; Owen, 1979, pp. 349–354; Pecori, 1981; Trinkaus, 1978). It is repeatedly emphasized by technical experts in this field that there is no complete substitute for an alert and observant individual and that visual means of detection coupled with judicious mechanical probing using the simplest of tools (sharpened poles, long-handled rakes) maintain their pre-eminent role in many instances (Greene, 1969; Lumsden, 1978, p. 194; Quinn, 1971; Thanh, 1974).

One method of munition detection that seems to offer substantial promise is the use of specially trained dogs for the purpose (Kelch, 1982; Lubow, 1977, pp. 173–202; Quinn, 1971; see also chapter 7). Dogs have been used in this way in the past and, although considerable skill and effort must go into their training, the approach should be explored with vigour. The olfactory abilities of other animals might prove useful as well (Meyer, 1982).

Military forces rely on a number of techniques for breaching or overcoming minefields beyond those just mentioned that, however, appear to have at best only limited civil applicability (Army, 1984, pp. 74–75; Foss & Gander, 1984, pp. 238–249; Hughes, 1979; Kitching, 1977; Watson, 1961; Zhuravlyov, 1974). These include most prominently heavy rollers or crushers mounted on the front of a vehicle which are meant to clear a path through a minefield, or else a long thin charge which is pushed or propelled ahead and set off to accomplish the same thing. It has been suggested by some that fuel–air explosive munitions can be detonated in order to clear the terrain below the blast (Dennis, 1976). It has even been seriously suggested to spray a fast-curing foam plastic path on top of a minefield so that soldiers can walk across it in safety, employing the load-distribution principle of the snowshoe (Marsden, 1975). One method that can be moderately useful under some circumstances is to set a ground fire (assuming that fuel, wind, and other conditions are appropriate) and let it sweep over a minefield.

Once detected, a mine or other explosive munition must be neutralized. This is most readily accomplished by blowing it up in place. If local circumstances do not permit this, the more dangerous method of dismantling (defusing) the device is called for. Expert knowledge is required in either instance, especially the latter.

Disposal at sea

The neutralization of sea mines involves quite sophisticated equipment for detection, including specially equipped ships (minesweepers) and, more recently, helicopters (Davis, 1978; Marriott, 1974–1975; McCauley, 1974; Wettern, 1979; Williams, 1979; see also chapter 6). Mechanical, acoustic, magnetic and other sensors are employed. Sea mines are often designed so as to minimize their detectability, and the same area must be swept many times before it is considered safe. Depth charges will set off many, but not all kinds of mines. In shallow coastal and harbour areas it is often considered necessary to supplement ship and helicopter sweeping with detailed underwater searches by divers. Sea-mine clearing promises to remain an extraordinarily difficult task (McCoy, 1975; Truver, 1985).

V. Conclusion

The long-term post-war problem of material remnants of war, particularly of mines and other unexploded ordnance, is a singularly grave and intractable one. This is the case especially in view of the following factors: (*a*) the great—and, to a substantial extent, successful—efforts in mine design and emplacement techniques meant precisely to prevent their discovery by the enemy; (*b*) the efforts to keep munitions functional in the face of extremely adverse environmental conditions (quite successful, to judge from their extraordinary longevity); (*c*) the vast and ever growing levels of wartime munition expenditures, exacerbated by such recent ordnance innovations as cluster bomb units and scatterable mines; and (*d*) the growing emphasis, at least in some types of warfare, on a strategy of large-scale area neutralization or area denial. A corollary of factor (*a*) above is the reticence of the countries that employ sophisticated land or sea mines to disclose those technical details that would be useful in devising suitable detection devices. And a corollary of factors (*c*) and (*d*) above is the increasing difficulty in keeping track of where mines have been distributed.

The problem of explosive remnants could be mitigated through the design and adoption of more dependable fuses that would thus result in the generation of a smaller residuum of dangerous duds. Moreover, every type of high-explosive ordnance—mines and all the rest—should be designed to become harmless in due course following its expenditure. Existing detection equipment should be refined and new types developed, especially with reference to non-metallic detection. An open clearing house and repository should be established for relevant information, presumably under United Nations (possibly UNEP) auspices, perhaps in conjunction with an international research, training and information centre. Such a centre should include a department in the use of dogs for detection on land and possibly an equivalent one in the use of marine mammals for underwater detection.

The problem of explosive remnants could be mitigated by more widespread adoption and adherence by the nations of the world to the relevant multilateral treaties (see chapter 8). The first of these is the Hague Convention VIII of 1907 (see appendix 5), which places a number of restrictions on the use of sea mines (Levie, 1971–1972; Sandoz, 1981). The treaty entered into force in 1910 and has only 27 parties as of January 1985 (although including China, France, the United Kingdom and the USA from among the five permanent members of the United Nations Security Council). Secondly there is the Inhumane Weapon Convention of 1981 (see appendix 3), Protocol II of which provides a lengthy series of restrictions on the use of land mines and booby traps (Fenrick, 1981). The treaty entered into force in 1983 and as of January 1985 there are about 22 nations that have ratified it (including China and the USSR from among the five permanent members of the United Nations Security Council). Relevant to the problem in a broader sense are those multilateral treaties meant to exclude *any* military activity from an area (e.g., the Spitsbergen Treaty of 1920, the Aaland Convention of 1921, and the Antarctic Treaty of 1959) as well as those meant to prevent certain general classes of especially obnoxious military behaviour or activity (e.g., Hague Convention IV on the Laws and Customs of Land War of 1907, Geneva Convention IV on the Protection of Civilians in War of 1949 plus its Additional Protocol I of 1977, and the Environmental Modification Convention of 1977).

The problem of explosive remnants could be mitigated by greater co-operation among nations including an open exchange of information, joint research efforts and multinational clean-up programmes. The initiative of the United Nations General Assembly in 1975 is to be lauded in recognizing that certain developing countries are being impeded in their development by material remnants of war, especially mines, in requesting co-operation from the responsible states, and in calling upon the United Nations Environment Programme to undertake a study (UNGA, 1975; see also appendix 2). The United Nations Environment Programme has, in fact, initiated such a study, which includes a detailed questionnaire sent to all countries, and two initial reports are available (Tolba, 1977; appendix 8).

In closing it must be stressed that the impact of explosive remnants of war on the human environment demands attention because of the tragic losses of life a errible maimings that continue to occur even long after the cessation of ho .: ies. These remnants also kill livestock and wildlife. And, of course, they impede post-war reconstruction and rehabilitation efforts particularly in regard to agricultural, forestry, fishing, mining and other rural pursuits. The

existing legacy of our past wars is an immense one and the future bodes
worse. Finally, we must never lose sight of the simple fact that an end to war
would in time bring an end to this terrible problem of the explosive remnants
of war.

References

Alder, K. 1980. Modern land mine warfare. *Armada International*, Zurich, **4** (6): 6–18.

Anderberg, B. 1981. *Communication:* Geneva: UN Inst. for Training & Research Publ.
No. UNITAR/EUR/81/WR/20, 5 pp.

Andrews, M. A. 1978. Tank-delivered scatterable mines. *Military Review*, Ft
Leavenworth, Kansas, **58**(12): 34–39.

Army, US. 1984. *1984 weapon systems*. Washington: US Department of the Army, 137
pp.

Associated Press. 1974. Old bomb kills 3 in Japan. *New York Times* **1974** (3 Mar): 13.

Associated Press. 1977a. Bomb hunt ends in the waterways of West Berlin. *International Herald Tribune*, Paris, **1977** (26–27 Mar): 5.

Associated Press. 1977b. Explosives from war still kill Vietnamese. *International Herald Tribune*, Paris, **1977** (18–19 Jun): 3.

Associated Press. 1979. 8 Lebanese children killed as a live bomb explodes. *New York Times* **1979** (13 May): 11.

Associated Press. 1981. 5 French children die as old shell explodes. *New York Times*
1981 (20 Mar): A14.

Associated Press. 1982a. Despite mines, Falklanders cut peat for fuel. *International Herald Tribune*, Paris, **1982** (15 Dec): 5.

Associated Press. 1982b. Hills around Falkland capital strewn with Argentine dead.
New York Times **1982** (9 Jul): A3.

Associated Press. 1982c. Over 250 kinds of explosives found in Beirut. *New York Times*
1982 (12 Dec): 13.

Associated Press. 1984. Relics of World War still explode in Europe. *New York Times*
1984 (2 Dec): 24.

Berry, F. C., Jr. 1979. US Navy mine warfare: small but not forgotten. *Armed Forces
Journal International*, Washington, **117**(2): 38–39, 43.

Boyd, J. H. 1976. Nimrod Spar: clearing the Suez Canal. *United States Naval Institute
Proceedings*, Annapolis, Maryland, **102**(2): 18–26.

Brendt, W. 1967. Danger: booby traps. *Infantry*, Ft Benning, Georgia, **57**(3): 42–43.

British Ministry of Defence. 1980. *British defence equipment catalogue*. 12th ed. Farn-
borough, UK: Combined Service Publications, 1538 + 333 pp.

Cestac, R. 1981. *Technical aspects related to material remnants of war in Libyan battle-
fields*. Geneva: UN Inst. for Training & Research Publ. No. UNITAR/
EUR/81/WR/3, 35 pp.

Ceva, L. 1981. *Influence of mines and minefields in the north African campaign of
1940–1943*. Geneva: UN Inst. for Training & Research Publ. No. UNITAR/
EUR/81/WR/2, 25 pp.

Chase, M. B. 1980. Unique new capability: scatterable mines. *Army R, D & A*,
Alexandria, Virginia, **21**(2): 6–9.

Courson, M. 1972. Kahoolawe: island of death. *American Forests*, Washington, **78**(5):
16–19.

Crèvecoeur, P. 1977. FFV 028 anti-tank mine. *International Defense Review*, Geneva,
10: 529–530.

Davis, R. M. 1978. Helicopter operations: minesweeping. *International Defense Review*,
Geneva, **11**: 385–388.

Dennis, J. A. 1976. SLUFAE: long-range minefield breaching system tested. *Army Research and Development* [now *Army R, D & A*], Alexandria, Virginia, **17**(3): 14–15.

Evangelista, M. A. 1982–1983. Stalin's postwar army reappraised. *International Security*, Cambridge, Massachusetts, **7**(3): 110–138.

Fenrick, W. J. 1981. New developments in the law concerning the use of conventional weapons in armed conflict. *Canadian Yearbook of International Law*, Vancouver, **19**: 229–256.

Feron, J. 1982. Unease invades Falklands in war aftermath. *New York Times* **1982** (26 Jul): A1–A2.

Finnish Defence Forces. 1976. *[Defence Force activities, 1976.]* (In Finnish). Helsinki: Defence Headquarters, Depts of Planning & Information, 48 pp. + 5 apps.

Fonda, J. 1974. Rebirth of a nation: a Vietnam journal. *Rolling Stone*, San Francisco, **1974**(164): 48–58.

Foss, C. F. 1979. Mines in land warfare: a *Defence* survey. *Defence*, Eton, UK, **10**: 231–235.

Foss, C. F. (ed.). 1984–1985. *Jane's armour and artillery*. 5th ed. London: Jane's Publishing Co., 897 pp.

Foss, C. F. & Gander, T. J. (eds). 1984. *Jane's military vehicles and ground support equipment*. 5th ed. London: Jane's Publishing Co., 871 pp.

Goad, K. J. W. & Halsey, D. H. J. 1982. *Ammunition (including grenades and mines)*. Oxford: Brassey's Publishers, 289 pp.

Golino, L. 1984. Land mines: detection and countermeasures. *Defence Today*, Rome, **8**: 318–326.

Graves, W. 1975. New life for the troubled Suez Canal. *National Geographic Magazine*, Washington, **147**: 792–817.

Greene, W. M., III. 1969. Countermeasures against mines and booby traps. *Marine Corps Gazette*, Quantico, Virginia, **53**(12): 31–37.

Halloran, B. F. 1972. Soviet land mine warfare. *Military Engineer*, Washington, **64**: 115–118.

Hartmann, G. K. 1979. *Weapons that wait: mine warfare in the U.S. Navy*. Annapolis, Maryland: Naval Institute Press, 294 pp.

Hickey, G. C. 1973. Can South Vietnam make it on its own? *U.S. News & World Report*, Washington, **75**(7): 55–56.

Hiebert, M. & Hiebert, L. 1978a. Indochina war's effects linger on. *Guardian*, London, **1978** (6 Apr): 9.

Hiebert, M. & Hiebert, L. 1978b. New occupational hazard: unexploded bombs. *Guardian*, London, **1978** (7 Apr): 7.

Hill, J. R. M. 1984. Sappers in the Falklands. *Military Engineer*, Alexandria, Virginia, **76**: 162–167.

Hogg, I. V. 1981. Land mine technology today. *Defence*, Eton, UK, **12**: 182–187.

Hogg, I. V. (ed.). 1984–1985. *Jane's infantry weapons*. 10th ed. London: Jane's Publishing Co., 957 pp.

Horne, C. F., III. 1982. New role for mine warfare. *United States Naval Institute Proceedings*, Annapolis, Maryland, **108**(11): 34–40.

Howell, M. L., Jr. 1977. Scatterable mines. *Military Engineer*, Washington, **69**: 396–399.

Hughes, B. C. 1979. Combat engineering equipment for the 1980's. *Military Engineer*, Washington, **71**: 406–410.

Hyman, A. 1982. Bombs and unguided rockets: low-cost ordnance for aerial warfare. *Military Technology*, Bonn, **6**(4): 55–64.

Kelch, W. J. 1982. Canine soldiers. *Military Review*, Ft Leavenworth, Kansas, **62**(10): 33–41.

Kelly, J. 1983. Falkland Islands: a melancholy anniversary. *Time*, New York, **121**(13): 38–39.

Kitching, J. 1975. Land mine warfare. *International Defense Review*, Geneva, **8**: 691–694.

Kitching, J. 1977. Minefield breaching. *International Defense Review*, Geneva, **10**: 523–525.

Krepon, M. 1973–1974. Weapons potentially inhumane: the case of cluster bombs. *Foreign Affairs*, New York, **52**: 595–611.

Lancesseur, B. 1985. Naval mines. *Defence & Armament*, Paris, **1985**(38): 15–17.

LeBarron, R. K. & Walker, R. L. 1971. Kahoolawe. *Aloha Ainu*, Honolulu, **2**(2): 16–20.

Levie, H. S. 1971–1972. Mine warfare and international law. *Naval War College Review*, Newport, Rhode Island, **24**(8): 27–35.

Libya. 1981. *White book: some examples of the damages caused by the belligerents of the World War II to the people of the Jamahiriya*. Tripoli: Libyan Studies Centre, 176 pp.

Lowe, J. R. 1968. EOD in Vietnam. *Ordnance*, Washington, **53**(289): 71–74.

Lubow, R. E. 1977. *War animals*. Garden City, NY: Doubleday, 255 pp.

Luce, D. 1974. Vietnam eyewitness: rebuilding underway. *Guardian*, New York, **26**(15): 13.

Luckow, U. 1982. Victory over ignorance and fear: the U.S. minelaying attack on North Vietnam. *Naval War College Review*, Newport, Rhode Island, **35**(1): 17–27.

Lumsden, M. 1978. *Anti-personnel weapons*. London: Taylor & Francis, 299 pp. [a SIPRI book].

Mahon, G. H. (ed.). 1972. *Department of Defense appropriations for 1973*. Washington: US House of Representatives Committee on Appropriations, Pt 4, 1248 pp.

Marriott, J. 1974–1975. Mine warfare. *Nato's Fifteen Nations*, Amstelveen, **19**(5): 44–50.

Marriott, J. 1975–1976. Modern tactical air-to-ground weapons. *Nato's Fifteen Nations*, Amstelveen, **20**(2): 40–49.

Marsden, J. N. 1975. Defeat of tactical mine fields. *National Defense*, Washington, **60**: 127–129.

Martin, E. 1973. Defusing the rice paddies. *Washington Post* **1973** (8 Jul): D6.

McCauley, B. 1974. Operation End Sweep. *United States Naval Institute Proceedings*, Annapolis, Maryland, **100**(3): 18–25.

McCoy, J. M. 1975. Mine countermeasures: who's fooling whom? *United States Naval Institute Proceedings*, Annapolis, Maryland, **101**(7): 39–43.

McDavitt, P. W. 1979. Scatterable mines: superweapon? *National Defense*, Arlington, Virginia, **64**(356): 33–37.

McWhirter, W. 1982. Falkland islands: saved but still fearful. *Time*, New York, **120**(6): 32–34.

Mercer, D. C. 1969. Mine-clearing methods: engineer mine sweepers. *Military Engineer*, Washington, **61**: 16–17.

Meyer, D. G. 1982. Sniffing out dangerous mines a real turn-on for rats. *Armed Forces Journal International*, Washington, **119**(7): 24.

Milling, J. S. 1969. Mines & booby traps. *Infantry*, Ft Benning, Georgia, **59**(1): 39–40.

New York Times. 1973. Admiral reports 11,000 mines dropped in the North's waters. *New York Times* **1973** (30 Mar): 17.

Niemann, K.-P. 1983. [Mine warfare: role in the concept of the Navy.] (In German). *Soldat und Technik*, Frankfurt a.M., **26**: 64–67.

Oliver, F. W. 1945–1946. Dust-storms in Egypt and their relation to the war period, as noted in Maryut, 1939–45. *Geographical Journal*, London, **106**: 26–49 +4 pl.; **108**: 221–226 + 1 pl.

Owen, J. (ed.). 1979. *Brassey's infantry weapons of the world*. 2nd ed. London: Brassey's Publishers, 480 pp.

Patterson, A. 1970–1971. Mining: a naval strategy. *Naval War College Review*, Newport, Rhode Island, **23**(9): 52–66.

Pecori, P. M. 1981. Army solves desert mine detector problem. *Army R, D & A*, Alexandria, Virginia, **22**(2): 18–19.

Pengelley, R. B. 1974. Clearing the Suez Canal. *International Defense Review*, Geneva, **7**: 735–736.

Pretty, R. T. (ed.). 1984–1985. *Jane's weapon systems*. 15th ed. London: Jane's Publishing Co., 1017 pp.

Prokosch, E. 1976. Antipersonnel weapons. *International Social Science Journal*, Paris, **28**: 341–358.

Quinn, W. L., Jr. 1971. Dogs in countermine warfare. *Infantry*, Ft Benning, Georgia, **61**(4): 16–18.

Red Cross, International Committee of the. 1973. *Weapons that may cause unnecessary suffering or have indiscriminate effects*. Geneva: Intl Committee of the Red Cross, 72 pp.

Reuters. 1975. Japan still hunts mines sown in '45. *New York Times* 1975 (5 Jan): 5.

Rouarch, C. 1984. Naval mine: as effective a weapon as ever. *International Defense Review*, Geneva, **17**: 1239–1240, 1245–1248.

Ryabchikov, V. 1977. Civil defence of the USSR. *Soviet Military Review*, Moscow, **1977**(2): 46–47.

Rybicki, J. F. 1984. Land mine warfare and conventional deterrence. *NATO's Sixteen Nations*, Brussels, **29**(5): 75–82.

Sandoz, Y. 1981. *Unlawful damage in armed conflicts and reparation therefor under international humanitarian law*. Geneva: UN Inst. for Training & Research Publ. No. UNITAR/EUR/81/WR/7, 35 pp.

Searle, W. F., Jr & Moody, D. H. 1981. *Clearance of explosive ordnance, sea mines and other war debris from the marine environment*. Geneva: UN Inst. for Training & Research Publ. No. UNITAR/EUR/81/WR/18, 35 pp.

Stampfer, R. 1981. [Mines.] (In German). *Soldat und Technik*, Frankfurt a.M., **24**: 80–88.

Studenikin, P. 1975. Gulf of Suez: on tack with courage. *Soviet Military Review*, Moscow, **1975**(2): 49–50.

Swearington, [T.] 1969. *Staff study on pernicious characteristics of U.S. explosive ordnance*. Washington: US Marine Corps, unpubl. ms (Oct 69), 10 pp.

Taylor, J. D. 1977. Mining: "a well reasoned and circumspect defense". *United States Naval Institute Proceedings*, Annapolis, Maryland, **103**(11): 40–45.

Thanh, Yen. 1974. Paddies on old no-man's-land. *South Viet Nam in Struggle*, Hanoi, **8**(242): 5.

Time. 1977. Return of the natives to Kahoolawe. *Time*, New York, **110**(6): 32.

Tolba, M. K. 1977. *Implementation of General Assembly resolution 3435 (XXX): study of the problem of the material remnants of wars, particularly mines, and their effect on the environment*. Nairobi: UN Environment Programme Document No. UNEP/GC/103 (19 Apr 1977), 8 pp. + UNEP/GC/103/Corr.1 (6 May 1977), 1 p. Also: New York: UN General Assembly Document No. A/32/137 (27 Jul 1977), 1 + 8 pp.

Tresckow, A.v. 1975. [Land mines.] (In German). *Soldat und Technik*, Frankfurt a.M., **18**: 388–400.

Trinkaus, H. P. 1978. METEX and FEREX state-of-the-art hand-held mine detectors. *International Defense Review*, Geneva, **11**: 1482–1484.

Truver, S. C. 1985. Mines of August: an international whodunit. *United States Naval Institute Proceedings*, Annapolis, Maryland, **111**(5): 94–117.

UNGA (United Nations General Assembly). 1975. *United Nations Environment Programme*. New York: United Nations General Assembly Resolution No. 3435

(XXX) (9 Dec 1975), 1 p. Reprinted in: *UN Yearbook*, New York, **29**: 443. Also in: *UN Disarmament Yearbook*, New York, **1**: 255.

United Press International. 1976. Old shell kills 21. *International Herald Tribune*, Paris, **1976** (6 Jul): 5.

United Press International. 1983. Falklands mine hurts U.K. major. *International Herald Tribune*, Paris, **1983** (18 Jan): 2.

Vehnekamp, B. 1970. Explosive ordnance disposal. *Military Engineer*, Washington, **62**: 308–309.

War, US Dept of. 1943. *Ordnance: unexploded bombs: organization and operation for disposal*. Washington: US Dept of War Field Manual No. FM 9–40, 148 pp.

Watson, M. S. 1961. Underground warfare. *Ordnance*, Washington, **45**: 770–774.

Westing, A. H. 1975. Unexploded munitions problem: an American legacy to Indochina that still wounds and kills. *Rutland [Vermont] Herald & Times Argus* **1975** (14 Dec, Pt IV): 3.

Westing, A. H. 1980. *Warfare in a fragile world: military impact on the human environment*. London: Taylor & Francis, 249 pp. [a SIPRI book].

Wettern, D. 1979. Mines. *Navy International*, Surrey, UK, **84**(3): 12–17.

Whelan, M. J. 1980. Soviet mine warfare: intent and capability. *United States Naval Institute Proceedings*, Annapolis, Maryland, **106**(9): 109–114.

Wilcke, J. 1971. [Underwater explosion: processes, consequences, problems.] (In German). *Soldat und Technik*, Frankfurt a.M., **14**: 74–77.

Wildrick, E. W., III. 1969. Mines and booby traps in Vietnam. *Military Engineer*, Washington, **61**: 7–8.

Wile, T. S. 1982. Their mine warfare capability. *United States Naval Institute Proceedings*, Annapolis, Maryland, **108**(10): 145–151.

Williams, R. N. 1979. Finding and clearing mines. *Armor*, Washington, **88**(6): 12–16.

Zhuravlyov, A. 1974. Breaching obstacles. *Soviet Military Review*, Moscow, **1974**(10): 42–43.

2. Explosive remnants of World War II in Poland[1]

Boguslaw A. Molski and Jan Pajak[2]
Polish Academy of Sciences

I. Introduction

Poland was in a state of conflict during World War II from September 1939 to May 1945, a period of fully five years and eight months (Topolski, 1981). These years of battle had a brutal impact on this country's land and people. Poland sustained an estimated six million fatalities during that time—some 17 per cent of the population and thus by far the highest proportion of any nation during World War II (Westing, 1980, p. 35). By the end of the war, much of Poland was heavily saturated with undetonated mines and other explosive munitions, as were its adjacent waters. Its post-war problem of explosive remnants may have been the most serious of any country involved in that war.

Following a brief explanation of the military background, this chapter describes the types of minefields and mines as well as the types of other explosive munitions (duds) that were present on the land of Poland after the war. It goes on to outline how Poland coped with its explosive-remnant problem from both the organizational and technical standpoints. Rural land clearance is discussed separately from urban clearance. A section on explosive remnants at sea follows. The chapter concludes with a brief indication of the societal costs of the explosive remnants of war and some suggestions for ameliorating the problem in the future.

[1] This chapter was abridged by the editor from the authors' much more detailed and heavily illustrated manuscript.
[2] The authors are pleased to acknowledge assistance from Zdzislaw Stelmaszuk, Andrzej Szerauc and Stanislaw Skiers of the Polish Ministry of National Defence and from Tadeusz Kulikowski of the Polish Ministry of Administration and Municipal Economy. Factual data presented in the text without specific citation are from the unpublished archives of the Polish National Ministry of Defence Army Combat Engineers.

II. Military background

The World War II military activities in which Poland was embroiled can be divided into three phases: (*a*) 1 September to 4 October 1939; (*b*) October 1939 to mid-July 1944; and (*c*) July 1944 to April 1945.

During the initial one-month campaign in September 1939 Germany conquered Poland. On the German side were 1.8 million soldiers with 11 000 artillery pieces, 2 800 tanks and 2 600 aircraft; and on the Polish side 1.2 million soldiers with 3 000 artillery pieces, 600 tanks and 400 aircraft. The fighting during this brief phase of the war was heavy, especially around Warsaw and a number of other locations. Large numbers of bombs as well as huge numbers of artillery and mortar shells were expended, leaving behind the inevitable fraction of duds; and some areas were mined with anti-tank and/or anti-personnel mines.

During the almost five-year period 1939–1944 a variety of organized and unorganized Polish underground forces—altogether numbering some half-million armed individuals—fought numerous brief campaigns and skirmishes against the German occupation forces in many parts of the country. As to the use of explosive munitions during this phase of the war, one must point especially to grenades and mortar shells.

During the final nine months, from July 1944 to April 1945, Poland became the major battlefield between attacking Soviet forces and defending German forces. This was the phase of the war during which Poland was so heavily shelled and mined by both of the contending armies. Many cities were declared as fortresses by Germany and were heavily fortified and mined, including Warsaw, Wroclaw (Breslau), Krakow, Gdansk (Danzig), Poznan (Posen), Opole (Appeln), and Olsztyn (Allenstein). The Germans also heavily fortified several rural regions in northern Poland. The lines became stabilized more or less along the Wisla (Vistula) River for some months during the latter part of 1944 and each of the two armies fortified and mined its side to a depth of a few kilometres. Then, during their advance in early 1945, the Soviets heavily mined southern Poland in order to help secure this flank from the enemy.

III. Explosive remnants on land

Mines

In early 1945 the USSR began providing Poland with documentation on the minefields it had laid, which in time included more than 17 000 maps. In 1947 the USA and the United Kingdom provided Poland with captured information on 1 500 German minefields. It could be determined from these and other data that the total land area of Poland which had been mined was about 25 million hectares in size, that is, covering about 80 per cent of the

Table 2.1. Land area mined in Poland during World War II

Intensity of mining (mines/hectare)	Area (10^6 hectares)	Proportion (per cent)
More than 10^a	0.6	*2*
1–10	1.4	*4*
Less than 1	23.0	*74*
Essentially 0	6.3	*20*
Total	**31.3**	***100***

[a]The value usually falling between 10 and 60.

Source: Pajak (1973).

country (table 2.1). Of this total, almost 600 000 hectares—2 per cent of the country—had been especially heavily mined.

The heavily mined areas could be categorized into some half-dozen distinct types on the basis not only of their terrain features (relief, water table, plant cover, etc.) and cultural features (towns, factories, bridges, railways, etc.), but also as to type of mine and minefield, extent of booby trapping, degree of subsequent military disruption by artillery shelling and the like, and number and variety of explosive dud munitions (Kaczmarski & Soroka, 1982). Mines were often laid in thick vegetation, in swamps, along river banks, in narrow passes, around towns, and, of course, around fortifications and along front lines. At different times and places, mines were variously employed defensively or offensively, as well as for diversion or sabotage. Some land areas changed hands several times; these were inevitably mined by both sides and disrupted by battle as well. Such zones presented especially hazardous conditions during post-war clearance operations.

In addition to the heavily mined areas, large portions of the country were left infested with scattered minefields and smaller groups of mines of many sorts, together with huge numbers of unexploded artillery and mortar shells, bombs and grenades. Indeed, since the end of the war, five times as many of these mortally dangerous duds have been recovered in Poland as mines (table 2.2).

In a more or less stable military situation, a German defensive line might take the following form (Kaczmarski & Soroka, 1982; Pajak, 1973). At the foremost edge there would be barbed-wire entanglements with anti-personnel mines. Directly behind these anti-personnel minefields were anti-tank minefields with interspersed anti-personnel mines. Some 25 metres rearward were the first line of trenches with three or four rows of trip-wire actuated anti-personnel mines behind them. Each of the forward anti-tank minefields was usually between 10 and 25 metres wide and 25 to 100 metres long, containing rows of mines 2 to 10 metres apart in which the mines were 1 to 2 metres apart. About half the mines were anti-personnel mines and of these about 90 per cent would be actuated by pressure and the remaining 10 per

Table 2.2. Explosive remnants of World War II neutralized on land in Poland during 1945–1982

Year	Land mines (10³)	Large bombs and shells[a] (10³)	Small shells and grenades[a] (10³)	Total bombs, shells and grenades (10³)	Total remnants (10³)
1945	10 240	18 993	3 468	22 461	32 701
1946	2 954	5 349	1 832	7 181	10 136
1947	1 198	2 944	1 466	4 410	5 608
1948	174	849	3 613	4 462	4 636
1949	26	192	2 133	2 326	2 351
1950	28	125	3 255	3 380	3 409
1951	12	105	2 510	2 615	2 628
1952	25	208	946	1 154	1 179
1953	41	148	1 584	1 732	1 773
1954	29	27	2 467	2 494	2 523
1955	11	103	3 055	3 158	3 169
1956	26	80	3 343	3 424	3 450
Subtotal (1945–56)	14 764	29 124	29 673	58 797	73 560
1957	18			2 369	2 387
1958	14			1 294	1 309
1959	15			647	662
1960	8			601	609
1961	6			428	434
1962	6			545	551
1963	3			338	342
1964	3			543	547
1965	3			834	837
1966	2			861	863
1967	2			472	475
1968	2			506	508
1969	2			661	663
1970	2			350	352
1971	3			390	393
1972	3			593	597
1973	1			472	474
1974	1			387	388
1975	12			466	478
1976	1			331	332
1977	1			470	471
1978	1			336	337
1979	7			216	222
1980	1			212	212
1981	10			205	215
1982	2			236	238
Subtotal (1957–82)	130			14 766	14 896
Total	**14 894**			**73 563**	**88 457**

[a]'Large' signifies a calibre (diameter) of 38 millimetres or greater; 'small' signifies a calibre of less than 38 millimetres.

Source: Polish Ministry of National Defence, Warsaw, Army Combat Engineer annual reports (unpublished archives).

cent by trip wire. Soviet minefields differed from the German ones in various ways. For example, they were often three or four times as wide, used different mine placement patterns, and had a mix of mines in which about two-thirds were anti-personnel.

The minefield documentation received by Poland usually included the location of the minefield indicated on a map having a scale of perhaps 1:1 000, the types and total numbers of mines laid, the nature of their fusing, a description of the mine-laying pattern, and perhaps other useful information. Documentation was, of course, not received for all minefields; some of the information received was incomplete or incorrect; there was often subsequent disruption from shelling and bombing; and there might have been subsequent changes in the soil (from flooding, etc.) or vegetation (growth, wind-thrown trees, etc.). Nevertheless, of the 15 million mines that have been neutralized since the war, fully 97 per cent could be cleared during the first three post-war years (table 2.2). Over the past two decades a few thousand scattered mines or small minefields have been discovered and neutralized each year as well as occasional large minefields, as in 1975, 1979 and 1981.

Numerous types of land mines were employed. German forces used 43 different kinds, including 19 types of anti-tank mine, 12 types of anti-personnel mine, and 12 special types (booby-trapped ones, those for use in rivers or against railway track, etc.). Soviet forces used 35 different kinds, including 10 anti-tank, 9 anti-personnel and 16 special types. A commonly encountered German anti-personnel mine was the model SMi35 equipped with either a pressure or trip-wire fuse and occasionally an additional anti-disturbance fuse. This is an especially dangerous metallic fragmentation mine because when actuated it jumps up about 1.5 metres before detonating. A similar Soviet model, OZM–152, was also common. Among the German and Soviet anti-tank mines were found both metallic and essentially non-metallic models. These were all pressure actuated blast mines, some of which were booby-trapped with anti-disturbance fuses. A variety of non-standard hand-made mines were also encountered (some of which had been emplaced to blow up bridges, buildings or railway lines), these often having originated from the various Polish underground forces. Booby traps, in which objects militarily or otherwise useful were designed to blow up when disturbed, were left behind in both rural and urban settings by the German forces when they retreated (Sobczak, 1963).

Duds

Of the bombs, artillery and mortar shells, grenades and other explosive munitions that were expended in such profligate numbers during the course of the fighting in Poland, an estimated 5 to 10 per cent did not explode at the time they were intended to. After the war these duds have continued to be found in huge numbers (table 2.2). Farmers turn them up in ploughing, peat diggers in digging peat, road crews in repairing roads, construction workers in excavating for cellars and children in their play activities.

There was, of course, no documentation for the vast numbers of potentially dangerous dud bombs, shells and grenades that remained after the war. These were not only present in far greater numbers than the mines, but became haphazardly hidden wherever battles occurred. Thus, of the 74 million such explosive remnants that have been neutralized to date less than half were cleared during the first three post-war years (table 2.2). And more than 200 000 per year continue to be found, with no clear end in sight. Hidden caches and other abandoned stores of explosive munitions also continue to be found from time to time.

The three most common causes of duds were: (*a*) defective fuses; (*b*) projectiles hitting the ground at such a narrow angle that the impact did not actuate the fuse; and (*c*) impact zones too soft to actuate the fuse, for example, a layer of snow, freshly tilled soil or boggy marshland.

Duds in which the fuse malfunctioned during the course of its action remain extremely dangerous since any jarring might complete the process and set off the main charge. Others in which the fuse did not begin its action might never go off or else might require a stronger shock to cause them to do so. Rusting or other deterioration of the fuse may in time make especially the latter category less liable to detonate. However all duds encountered must be treated as if they are in the extremely dangerous category.

IV.   Clearance on land

Rural clearance

Land clearance was from the outset divided into two parts: reconnaissance and neutralization. The reconnaissance crews worked separately from the neutralization crews (the sappers) that followed them. Reconnaissance was aimed not only at locating an area to be cleared, but also at estimating the nature and quantity of the explosive remnants and at determining the scope of the required neutralization activities. Expert planning was required for these two types of crews to work effectively on a co-ordinated and continuous basis.

Land clearance was carried out by specially trained units of the Polish Ministry of National Defence Army Combat Engineers (Kaczmarski & Soroka, 1982). Initial overall planning of a typical post-war land clearance operation was carried out at battalion level over a period of about 6–10 days. This would include the locating and fencing of minefields and the distributing of maps and other available documentation to the subordinate companies. The companies would do a more detailed reconnaissance of the minefields over the next 2–3 days, among other things, dividing up a minefield according to the type and distribution of its mines.

In the subsequent neutralization process, metallic mines, together with unexploded munitions, were often located with electronic metal detectors.

Pointed metal rods 3 to 4 metres in length were used as probes to locate both metallic and non-metallic mines and other explosive munitions. Their employment, though tedious, was generally the preferred method. These rods were used by gently poking the soil at an angle of 30° to 45° with the horizontal to a depth of 10 to 20 centimetres. Every obstruction thus encountered had to be carefully checked out by hand. Trained dogs were not available to the Polish sapper units.

Each sapper checked out a path about 2.5 metres wide for which he remained responsible, marking it out with broomsticks or similar stakes. Each mine located was flagged for subsequent neutralization by the same sapper who found it. A sapper always worked alone and at a distance of between 50 and 150 metres from any other person, the distance depending upon the types of mine in the minefield. If the minefield was known not to have been booby-trapped, the sapper would usually render a mine harmless by dismantling it. Otherwise it was blown up in place.

Mine detonation was accomplished in one of several ways. Pressure-actuated anti-personnel mines were set off by pulling a heavy roller across the minefield, and those actuated by the trip wires were set off by pulling special anchors across the field. One type of anchor in use plus its line could be shot out to a distance of 60 metres for subsequent manual retraction. The sappers would work from behind a portable steel shield. Anti-tank and mixed minefields presented a more difficult neutralization problem. Here a common method was to detonate each mine by the blast of a 5 kilogram charge suspended perhaps 125 centimetres above it. In some cases, a row of suspended 25 kilogram charges was set off prior to the detailed location process, which was then followed by the procedures already outlined. If the area being cleared was arable land, the first post-clearance ploughing was always performed by the sapper.

The clearance procedures outlined above were those basically used in Poland during the initial 12-year post-war period, that is, from 1945 to 1956. During this time of concerted clearance efforts 15 million mines and 59 million other explosive munitions were neutralized (table 2.2). As a result more than 700 000 hectares could be declared sufficiently safe to be returned to civilian uses, 400 000 hectares of which was arable land (table 2.3). This cleared land had contained about 100 explosive remnants per hectare, on average.

Immediately after the end of the war, Poland set in motion its national programme of clearance, with both the urban and rural operations beginning simultaneously. Outside the cities the country was divided into 11 clearance districts and 60 sub-districts. The full scope of the problem was not at first recognized. Indeed, early estimates of the total number of explosive remnants to be cleared turned out to be below the number neutralized in 1945 alone; and the initial number of sappers in service and other resources devoted to this activity had to be increased accordingly. By mid-1945 a force of about 10 000 military engineers was engaged in this activity (about half in urban

Table 2.3. Land cleared of explosive remnants of World War II in Poland during 1945–1956

Year	Arable land (10^3 ha)	Pasture land (10^3 ha)	Forest land (10^3 ha)	Total land (10^3 ha)	Remnants neutralized per hectare cleared[a]
1945–49	379.2	192.7	106.3	678.2	82
1950	5.9	1.4	2.4	9.7	351
1951	2.8	1.1	1.2	5.1	515
1952	1.9	0.7	2.0	4.7	251
1953	1.5	0.5	2.8	4.9	362
1954	0.7	0.4	7.0	8.0	315
1955	1.5	2.0	9.0	12.4	256
1956	4.9	2.4	14.0	21.3	162
Total	**398.4**	**201.2**	**144.7**	**744.3**	**99**

[a]Number of remnants neutralized (from table 2.2) divided by total land cleared.

Sources: Polish Ministry of National Defence, Warsaw, Army Combat Engineer annual reports (unpublished archives); Gatarz (1974).

operations and half in the countryside), a number that could be reduced in the subsequent years (table 2.4, note *a*). The job of the sapper—always a voluntary position—required not only considerable intelligence and the acquisition of exacting skills, but also fortitude, stamina, patience and self-control.

Rural clearing began with roads, railways, coal mines, factories, power stations, bridges and arable land. The entire civil service co-operated with the military service in gathering information on explosive remnants, in marking their locations and so forth.

An intensive nationwide programme of public education on explosive remnants was instituted which included the wide distribution of numerous types of instructional materials. This was necessary because of the appalling numbers of civilian casualties that resulted from these remnants, averaging over 900 annually during the early post-war years (table 2.5). A disproportionate number of these casualties were, and continue to be, children.

As much as possible had to be achieved in the immediate post-war period not only for reasons of safety and to permit the return of land to productive use, but also because the detection of the remnants becomes increasingly difficult with the passage of time (Bordzilowski, 1973–1974). Soil movement via wind and water, the fall of leaves and the growth of vegetation can readily obscure many originally telltale marks.

During 1945 the 10 000 Polish sappers in service were able to neutralize almost 33 million explosive remnants, including 10 million mines and 19 million large bombs and shells (table 2.2). They were thereby able to clear, or confirm the safety of, 25 million hectares (or 80 per cent) of the country,

Table 2.4. Effort involved in neutralizing explosive remnants of World War II on land in Poland during 1957–1982 (with a note on 1945–1956)

Year	Disposal personnel employed[a]	Time spent (10³ work-days)	Vehicular distance travelled (10³ km)	Explosives used (10³ kg)	Remnants neutralized per work-day[b]
1957	613	133	917	58	18
1958	570	87	777	40	15
1959	515	75	735	32	9
1960	344	60	805	35	10
1961	468	73	873	29	6
1962	400	60	741	37	9
1963	520	46	645	17	7
1964	520	45	743	18	12
1965	254	38	675	20	22
1966	300	47	790	16	18
1967	365	93	951	14	5
1968	365	89	889	14	6
1969	396	43	805	14	15
1970	408	36	781	10	10
1971	432	43	921	11	9
1972	478	53	1 143	14	11
1973	490	48	1 300	13	10
1974	436	47	1 492	11	8
1975	471	46	1 168	9	10
1976	449	42	1 113	11	8
1977	355	33	1 091	8	14
1978	378	41	1 042	7	8
1979	329	30	1 179	9	7
1980	366	30	1 155	13	7
1981	377	25	1 178	8	9
1982	368	34	1 128	7	7
Total		**1 397**	**25 037**	**475**	11

[a]Disposal personnel employed on land during the prior years were: 1945, 10 000; 1946, 4 750; 1947, 2 750; 1948, 2 250; 1949, 2 250; 1950, 3 750; 1951, 2 500; 1952, 3 500; 1953, 3 000; 1954, 2 500; 1955, 1 750; and 1956, 750.
[b]Number of remnants neutralized (from table 2.2) divided by time spent.

Source: Polish Ministry of National Defence, Warsaw, Army Combat Engineer annual reports (unpublished archives).

including 29 500 towns and villages, 170 000 kilometres of road, 5 020 bridges and 39 airports (Gatarz, 1974; Pajak, 1974).

Some extremely heavily mined areas—for example, Hitler's secret headquarters near Ketzyn (Rastenburg) (Kaczmarski & Soroka, 1982; Pluta, 1954), Dukla Pass in the Carpathian Mountains and the so-called 'Mortal Hill' in Gubin—simply had to be isolated for as long as a decade or even more following the war, until clearing could be accomplished. 'Mortal Hill' had changed hands several times during the war in very fierce fighting. When its 20 hectares were finally cleared more than 90 000 explosive remnants (mines, shells and booby traps) had to be neutralized.

Table 2.5. Casualities from explosive remnants of World War II neutralized on land in Poland during 1945–1981

Year	Civilians killed	Civilians wounded	Disposal personnel killed	Disposal personnel wounded	Disposal personnel killed per 10^6 remnants neutralized[a]
1945			301	473	9.2
1946			71	67	7.0
1947			46	39	8.2
1948	2 293	5 205	10	14	2.2
1949			4	4	1.7
1950			5	4	1.5
1951			0	2	0
1952			5	2	4.2
1953			2	9	1.1
1954	668	819	1	6	0.4
1955			3	9	0.9
1956			1	6	0.3
Subtotal (1945–56)	2 961	6 024	449	635	6.1[b]
1957	205	318	0	6	0
1958	77	108	0	0	0
1959	86	223	1	1	1.5
1960	85	195	1	2	1.6
1961	103	249	2	2	4.6
1962	74	169	3	4	5.4
1963	51	145	0	0	0
1964	68	142	0	0	0
1965	56	152	0	0	0
1966	33	130	0	0	0
1967	34	143	0	0	0
1968	41	118	0	0	0
1969	27	126	0	0	0
1970	32	70	0	0	0
1971	43	78	0	1	0
1972	20	87	2	2	3.4
1973	15	52	2	3	4.2
1974	15	50	0	0	0
1975	16	65	0	0	0
1976	6	13	0	0	0
1977	4	34	0	0	0
1978	10	30	0	0	0
1979	9	15	0	1	0
1980	10	11	0	0	0
1981	13	27	0	0	0
Subtotal (1957–81)	1 133	2 750	11	22	0.7[b]
Total	**4 094**	**8 774**	**460**	**657**	**5.2[b]**

[a] Number of disposal personnel killed divided by number of remnants neutralized (from table 2.2).
[b] Expressed in the reciprocal form, the number of explosive remnants neutralized per disposal personnel fatality were: 1945–56, 164 000; 1957–81, 1 350 000; and 1945–81, 192 000.

Source: Polish Ministry of National Defence, Warsaw, Army Combat Engineer annual reports (unpublished archives).

Today more than 300 disposal personnel respond to some 8 000 to 10 000 reports per year of explosive remnants being discovered. They must still neutralize over 200 000 remnants per year (table 2.2), collectively spending perhaps 30 000 work-days, travelling some one million kilometres and utilizing almost 10 000 kilograms of explosives to set off the remnants that must be blown in place (table 2.4). During a work-day each of the disposal personnel neutralizes an average of 11 explosive remnants.

Neutralizing explosive remnants is a hazardous occupation. Some 460 disposal personnel have been killed in the line of duty in Poland during the four decades following World War II and an additional 657 wounded (table 2.5). In other words, about five disposal personnel have been killed per million explosive remnants neutralized throughout this period (or, in reciprocal terms, 192 000 neutralizations have been accomplished per fatality). In fact, the fatality rate has dropped remarkably over the years. During the first 12 years (1945–56) the rate was about six fatalities per million neutralizations (or 164 000 neutralizations per fatality) whereas during the subsequent 26 years (1956–81) the rate dropped to less than one fatality per million neutralizations (or 1 350 000 neutralizations per fatality). This improvement can presumably be attributed in part to the fact that most mines were cleared during the initial 12-year period (see table 2.2) and in part to greater experience leading to improved techniques.[3]

Finally it must be emphasized again that the explosive remnants of World War II have resulted in thousands of civilian fatalities and injuries in Poland (table 2.5). Even at this late date, about a dozen people continue to be killed annually and twice that number injured. As noted earlier, these casualties include a disproportionately high number of children. To illustrate this point, of the 39 civilians killed during the recent five-year period 1976–80, 34 (or 87 per cent) were children; and of the 103 wounded, 84 (or 82 per cent) were children.

Urban clearance

Although major clearance operations had to be carried out in numerous urban areas, the biggest problem by far was presented by Warsaw. The liberation of Warsaw from the occupying German forces was a protracted and highly destructive affair. First Polish underground forces attempted an

[3] During the immediate two post-war years (1945–46) Poland cleared 43 million explosive remnants (13 million mines plus 30 million bombs, shells and grenades) (table 2.2), thereby incurring 912 disposal-personnel casualties (372 deaths plus 540 wounded) (table 2.5). This represents 21 casualties sustained per million explosive remnants neutralized. Although not directly comparable, it is interesting to note that during this same two-year period, France cleared 13 million mines, incurring 1 501 disposal-personnel casualties (115 casualties per million mines); Italy cleared 3 million mines, incurring 1 100 disposal-personnel casualties (367 casualties per million mines); and Belgium cleared 500 000 mines, incurring 286 disposal-personnel casualties (572 casualties per million mines) (Bordzilowski, 1973–1974). Over the 30-year period 1945–75 Austria cleared 46 000 mines, incurring 41 disposal-personnel casualties, 18 deaths plus 23 wounded (892 casualties per million mines) (unpublished letter from Austria to UNEP dated 25 November 1976). These data clearly support the suggestion that the casualty rate of disposal personnel drops as the number of neutralizations they carry out goes up.

uprising in the western sector of the city (i.e., west of the Wisla River) during August–September 1944, which was brutally crushed. Next, Soviet forces were able, in heavy fighting, to take control of the smaller eastern sector in September of that year, but it took until January to push the Germans out of the entire city. During that period the banks of the Wisla River were mined to an extraordinarily high level and in a chaotic fashion.

The Germans had earlier made the decision to destroy the whole of Warsaw before their departure. When the time came for them to carry out this plan, only the western sector of the city remained available to them. They set about to mine every building or other municipal structure (bridges, power stations, water works, sewer systems, monuments, etc.) that would not succumb to the torch. A large building, such as an apartment house, hotel, palace, library, church or hospital would be fitted with several huge strategically placed charges. These charges were hidden, booby-trapped and equipped for remote detonation. Before their withdrawal, the Germans were more than 80 per cent successful in burning or blowing up the city, that is to say, more than 80 per cent of Warsaw was converted to rubble.

The reconstruction of Warsaw became a matter of highest post-war priority for Poland (Jankowski & Ciborowski, 1978). But the problems to be faced were truly monumental. In addition to the vast amount of deliberate and incidental damage already alluded to, the city was saturated with explosive remnants of the war. On an area of less than 40 000 hectares, more than four million mines, bombs, shells, booby traps and other explosive munitions have to date been found and neutralized.

Polish sapper units began clearing the eastern sector of the city soon after its liberation. During the first three months they removed 11 200 mines and 9 700 bombs, shells and so forth. Once the western sector was also liberated, Polish military engineer units about 4 500 strong divided the city into 12 districts, each with its own disposal headquarters. The planning phase included the making of decisions as to which structures to clear and which to blow up as being too dangerous to clear. In the first stage (lasting about one month) the main streets were cleared (and their uncleared portions fenced off) and the major governmental, industrial, commercial and apartment structures rendered safe (or, if necessary, designated as being unsafe). In the second stage (lasting about two weeks) the remaining streets, the railways and airports, and the underground sewer system were cleared. In the final stage (lasting about three weeks) the major disposal work was completed throughout the city.

Clearing a building of explosive remnants was an extremely difficult task for various reasons, among them: that the mines were often in concealed locations; that they, and the building in general, were often extensively booby-trapped; that unexploded bombs and shells were often haphazardly lodged in the higher floors; that the building was usually without electricity; that its cellar might be flooded; and that the building might have sustained partial battle damage and be obstructed both inside and out with piles of rubble. For example, a bomb weighing several hundred kilograms would have to be extricated from its lodging in some upper storey, placed on a toboggan made of mattresses, slowly lowered down the stairs made smooth with planking and then carried out of the city to be safely detonated.

During the intensive initial three-month period of clearing in Warsaw, some two million explosive remnants were found and neutralized, which thus represents about half the total that has so far been found in the city. Cleared during those initial three months were 960 kilometres of streets, 200 government buildings, 3 350 apartment buildings, 90 kilometres of sewer tunnel, and 710 hectares of public parks and gardens. The munitions neutralized consisted of 15 000 mines, 3 000 bombs, 743 000 large shells, 1 318 000 small shells and grenades, and 66 000 kilograms of explosive materials which had been emplaced to destroy municipal structures (Pilinski, 1953).

A special disposal unit has continued to function in Warsaw since that time and each year neutralizes 30 000 or more explosive munitions, that is, an average of over 80 per day! This number thus represents about 5 per cent of what continues to be found annually throughout Poland (table 2.2).

V.   Explosive remnants at sea and their clearance

Navigation in the Baltic Sea was very dangerous following World War II. During the first two post-war years (from mid-1945 to mid-1947) 37 ships were blown up there and 21 seriously damaged owing to sea mines. An international sea-mine disposal commission was established immediately following World War II under the leadership of the USSR for the purpose of clearing the Baltic, Black and Barents Seas. The commission estimated that the Baltic contained some 85 000 sea mines.

Each of the Baltic nations was given the responsibility by the commission to clear a sector of the sea adjacent to it. Poland was assigned a marine sector north of its coastline about 36 000 square kilometres in size and containing an estimated 2 500 sea mines in an inshore area of 5 000 square kilometres (Soroka, 1983). During the war the United Kingdom had dropped 1 070 seabed influence mines into the Gulf of Gdansk, which further contained a total of 12 minefields that had been emplaced by Germany and the USSR. Gdansk and Gdynia harbours within the Gulf had been mined as well. The Gulf of Pomerania (Pomorska) had been mined by the United Kingdom with 860 seabed influence mines, and the fairway off it from Swinoujscie (Swinemünde) south to Szczecin (Stettin) by Germany with an additional 24 seabed influence mines. Hostile mining had also been carried out in the harbour of Kolobrzeg (Kolberg) and of several other port cities.

The USSR carried out a minesweeping operation on behalf of Poland from May 1945 to September 1947, providing several minesweepers, a crane-equipped ship for clearing shipwrecks, and a number of mine-disposal divers

(Soroka, 1983). It began by sweeping the Gulf of Gdansk, where it cleared the most important fairways of Gdansk and Gdynia harbours, including a total of 700 square kilometres of the Gulf. It cleared a further 900 square kilometres in the Gulf of Pomerania and around Kolobrzeg harbour. Altogether it destroyed 51 seabed mines, 16 drifting mines and a number of moored mines. During the operation two of the Soviet minesweepers struck mines, with one lost in this way in June 1945 together with its entire crew of 16 and the other severely damaged in August 1945.

In 1945 Poland re-acquired three small ships which it had lost to Germany during the war and refitted them as minesweepers. In 1946 it acquired nine minesweepers from the USSR. These 12 ships were made ready for service and their crews trained later in 1946 for the Polish Navy to begin a 27-year clearing operation in the Gulfs of Gdansk and Pomerania and elsewhere along the coast. By 1948 Poland had purchased or built an additional ten small ships that were pressed into minesweeping service. During the first ten years of the operation Poland was able to clear about 2 300 square kilometres, which still left about 2 700 square kilometres closed to navigation and fishing. During 1965–70 the Gdansk area was re-swept more intensively than before in preparation for the enlargement and rebuilding of Gdansk harbour so it could accept ships of much deeper draught. This involved the sweeping of 11 fairways and roadsteads with a combined area of 1 400 square kilometres.

The Polish sea-mine clearing operation could finally be declared finished in 1973. At that time it was decided that the World War II seabed influence mines had become nonfunctional as influence mines and were thus no longer a hazard to surface shipping. (They, of course, remain hazardous to any seabed operations.) It was further decided that the moored contact mines—which must be considered to have remained hazardous, but not many of which had been emplaced in the Polish sector during the war—had been detected and neutralized in adequate numbers by the sweeping operations.

VI.   The cost to society

The direct human costs of Poland's explosive remnants of World War II in terms of the many lives lost and injuries sustained has already been alluded to (see table 2.5). The level of resources that have had to be divered to the clearance operations has been indicated earlier as well (see table 2.4). A few of the additional societal costs of explosive remnants are suggested below.

The two million hectares of Poland that had an average of at least one mine per hectare (table 2.1) contained many thousands of kilometres of roads and railways, thousands of bridges, and some 2 000 factories that could not be used for greater or lesser periods of time ranging from months to years.

Perhaps 400 000 hectares of arable land plus 200 000 hectares of pasture land were unavailable for the months to years it took to render those lands safe again (table 2.3). The present authors estimate that this situation reduced Poland's total agricultural production during the five-year period immediately following the war by approximately 10 per cent. Loss in cereal production during the 12-year period 1945–56 may have amounted to a total of 2 100 million kilograms, and loss in meat production to a total of 122 million kilograms. Moreover, timber utilization was held up on the 145 thousand hectares of mined forest land until they were cleared over the first 12 post-war years (table 2.3).

The use of Poland's harbours and coastal waters was held up for the months and years it took to clear them sufficiently, thereby inhibiting maritime commerce and ocean fishing.

And one must note again the enormous added difficulty that explosive remnants caused in the reconstruction of Warsaw, Wroclaw and other major cities.

VII. Conclusion

Poland was perhaps more heavily afflicted by mines and other explosive remnants of World War II than any other country. This terrible legacy of war is still being felt 40 years after the cessation of hostilities despite a large and ever more efficient disposal operation that has been in continuous operation since the end of the war.

It can only be hoped that in the future no country will be subjected to a similar burden. However, until such time that explosive munitions are no longer used, the experience of Poland has amply demonstrated the need for belligerents to keep accurate records of their minefields, whether land or sea, and to hand them over immediately following the cessation of hostilities, together with technical details of the emplaced mines. In addition, all land and sea mines should be constructed so that they inactivate themselves after a set time. Perhaps bombs, shells, grenades and other explosive munitions could also be manufactured with a similar self-neutralization mechanism.

There is no doubt that rapid attention to the explosive remnants of war is enormously advantageous. It is therefore urged that international help be extended quickly after a war to those nations requesting it.

References

Bordzilowski, J. 1973–1974. [War has ended, the fight continues.] (In Polish). *Wojskowy Przeglad Historyczny*, Warsaw, **1973**(3): 538–576; **1974**(1): 160–192.

Gatarz, H. 1974. [Thirty years of fighting with rusty death.] (In Polish). *Przeglad Wojsk Ladowych*, Warsaw, **1974**(1): 9–17.

Jankowski, S. & Ciborowski, A. 1978. *Warsaw: 1945, today and tomorrow*. Warsaw: Interpress, 210 pp.

Kaczmarski, F. & Soroka, S. 1982. *[Combat engineers of the Polish Army, 1945–1979.]* (In Polish). Warsaw: Ministry of National Defence Publishers, 380 pp.

Pajak, J. 1973. *[Combat engineers in the protection of the natural environment of man.]* (In Polish). Katowice: Center of Technical Progress, Series in Problems of the Natural Environment, 61 pp.

Pajak, J. 1974. *[Combat engineers in the reconstruction and development of the Polish People's Republic during 30 years.]* (In Polish). Warsaw: Ministry of National Defence Publishers, Publ. No. 8—SP, pp. 28–74.

Pilinski, W. 1953. *[Polish sappers.]* (In Polish). Warsaw: Ministry of National Defence Publishers, 128 pp.

Pluta, M. 1954. [Mine neutralization at the former Hitler headquarters.] (In Polish). *Przeglad Inzynieryjny*, Warsaw, **1954**(3): 49–58.

Sobczak, K. 1963. [Main period of mine neutralization in Warsaw.] (In Polish). *Roczniki Warszawski*, Warsaw, **4**: 218–258.

Soroka, M. 1983. [Mine in the trawl.] (In Polish). *Zolnierz Polski*, Warsaw, **1983**(26): 12–13.

Topolski, J. (ed.). 1981. *[History of Poland.]* (In Polish). Warsaw: Polish Scientific Publishing House, 956 pp.

Westing, A. H. 1980. *Warfare in a fragile world: military impact on the human environment.* London: Taylor & Francis, 249 pp. [a SIPRI book].

3. Explosive remnants of World War II in Libya: impact on agricultural development

Khairi Sgaier[1]

University of Alfateh, Tripoli

I. Introduction

Libya has supported a largely agrarian society, with dry farming the main activity and major source of livelihood for the majority of its people. The prevailing climate—low and erratic rainfall, high temperatures, and hot dry summer winds—has led to an emphasis on open-range livestock raising and the growing of cereals (grains) dependent upon ambient rainfall.

World War II was extraordinarily disruptive of Libya. And it was often the cereal lands and rangelands so crucial to the Libyan citizen and the Libyan economy that served as the scene of battle. Munitions, and especially land mines, were lavishly employed during these North African campaigns. The purpose of the present chapter is to provide a brief preliminary examination of the impact of the explosive remnants of that war on Libyan agriculture.

II. Nature and extent of the problem

During World War II Libya was a colony of Italy and was itself not a direct party to the conflict. Nevertheless, a series of major battles were fought on Libyan territory during the period 1940–1943, largely between German and British forces (Libya, 1981, chap. 4; Mostofi, 1983, chap. 2). Immense amounts of bombing and shelling occurred (Libya, 1981, chap. 5) and huge numbers of land mines (both anti-personnel and anti-tank) were emplaced (Libya, 1981, chap. 6). Indeed, according to one perhaps conservative estimate, the total number of mines emplaced in Libya was about 5 million (Cestac, 1981, p. 12). More than 70 major minefields were established as well as innumerable lesser ones. Their locations in many instances have never

---

[1] This chapter was adapted by the editor from the author's symposium presentation.

been revealed by the forces which emplaced them. As a result thousands of Libyans have been killed or maimed since 1940, and the number keeps growing (Libya, 1981, pp. 146–147, 154–155).

III. Rural impact

The 176 million hectares of Libya can for present purposes be divided into three major areas, to a great extent determined by amount of rainfall received: cereal land, rangeland and arid land. Each was more or less seriously and widely affected by explosive remnants, but only the first two are singled out below. Even though the cereal and rangelands represent only five or six per cent of the total land area of Libya, they of course support most of the nation's vital agricultural economy.

Cereal (grain) lands

The cereal lands of Libya, which cover about 2.4 million hectares, have been traditionally devoted largely to two species: primarily barley and, secondarily, wheat. Some 183 000 hectares, or eight per cent, of these cereal lands were rendered unusable by explosive remnants of World War II (table 3.1).

Until clearing began, these unusable lands represented a total loss of 61 000 tonnes of grain per annum, destined primarily for human consumption, plus an additional total annual loss of 124 000 tonnes of straw, destined for livestock feed (table 3.2).

The explosive remnants in the cereal lands built up to a maximum between 1940 and 1943. A modest effort to rid these lands of their explosive remnants began at that time, became concerted in 1945, and was largely successful by 1972. A crude estimate of the total cereal loss is thus 16 times the original annual loss, that is, perhaps 980 000 tonnes of grain plus almost 2 million tonnes of straw.

Rangelands

The 3.2 million hectares of open rangeland in Libya—supporting such livestock as sheep, goats, camels and cattle—provide its people and economy

Table 3.1. **Libyan cereal (grain) lands rendered unusable by explosive remnants of World War II**

Cereal	Total area (10³ ha)	Unusable area (10³ ha)	Unusable area (per cent)
Barley	1 500	137	9
Wheat	900	46	5
Total	**2 400**	**183**	**8**

Source: Robb (1945).

Table 3.2. **Libyan cereal (grain) losses owing to explosive remnants of World War II**

Cereal	Annual yield (kg/ha)	Area mined (10^3 ha)	Annual loss (10^3 t)
Barley			
Grain	320	137	44
Straw	750	137	103
Wheat			
Grain	370	46	17
Straw	450	46	21
Total			
Grain		**183**	**61**
Straw		**183**	**124**

Source: Robb (1945).

with meat and milk, hide and hair (including wool), draught power and transportation. The explosive remnants of World War II have disrupted this resource both by killing livestock and by denying areas to grazing.

Almost 2.8 million hectares, or fully 87 per cent, of the Libyan rangelands were rendered unusable during World War II, and remained so, by virtue of having been mined or suspected of having been mined (table 3.3). Until these unusable lands were either cleared of explosive remnants or determined to be safe, their non-use represented a total loss of more than 100 million so-called feed units per annum, that is, forage for ruminants approximately equivalent to 100 000 tonnes of barley feed for a period ranging from 3 to 33 years, and in some areas to this day (table 3.4). As with the cereal lands, clearing operations have also progressed in the rangelands, but by no means as quickly, thoroughly or widely. By 1980 only about 1.8 million hectares, or 67 per cent, could be declared safe, and thus the tedious and dangerous clearing efforts will have to continue for years. A crude estimate of total rangeland loss to date is thus 22 times the original annual loss, that is, natural forage equivalent to over 2 million tonnes of barley feed.

Table 3.3. **Libyan rangelands rendered unusable by the presence or suspected presence of explosive remnants of World War II**

Rainfall (mm/yr)	Total area (10^3 ha)	Area mined (10^3 ha)	Area mined (per cent)
50–100	1 850	1 660	*90*
100–150	1 110	1 060	*95*
150–200	100	20	*20*
>200	130	20	*15*
Total	**3 190**	**2 760**	*87*

Sources: Gintzberger & Bayoumi (1972); Le Houérou (1965).

Table 3.4.   Libyan rangeland losses owing to the presence or suspected presence of explosive remnants of World War II

Rainfall (mm/yr)	Productivity[a] (feed units)	Area mined (10^3 ha)	Annual loss (10^6 feed units)
50–100	20	1 660	33
100–150	60	1 060	64
150–200	80	20	2
>200	125	20	3
Total		**2 760**	**102**

[a] Rangeland productivity is expressed in 'feed' units, each of which represents forage for ruminants equivalent, in terms of barley feed, to an annual yield of 1 kilogram per hectare; or in terms of feed energy, to an annual yield of 6.9 megajoules per hectare (Le Houérou & Hoste, 1977).

Sources: Gintzberger & Bayoumi (1972); Le Houérou (1965).

Table 3.5.   Libyan livestock losses owing to explosive remnants of World War II, during 1940–1980

Livestock	Approximate population in any year (10^3)	Total 40-year loss[a] (10^3)	Average annual loss[b] (number)	Average annual loss[b] (per 10^4)
Camels	250	75	1 900	75
Sheep	1 800	36	900	5
Goats	500	13	300	6
Cattle	250	1	25	1
Total	**2 800**	**125**	**3 100**	**11**

[a] The total 40-year loss values are underestimates owing to incomplete reporting to the authorities by the agricultural population, especially toward the beginning of the period.
[b] The average annual loss values represent 40-year averages which do not reflect that the recorded values were somewhat higher than average toward the beginning of the period and somewhat lower toward the end.

Source: Libyan Ministry of the Interior, Tripoli (unpublished archives).

Direct losses of livestock on the open range owing to explosive remnants continue to this day, the number of such deaths—averaged over the four decades since 1940—having been more than 3 000 per year, or 1 per thousand annually of the livestock population extant at a given time (table 3.5). A disproportionately high number of camels have been killed in this way because camels are traditionally permitted to move more freely and widely than the other livestock.

Further considerations

A number of losses to the agricultural sector must be noted beyond those directly related to cereals or livestock. The inevitable losses of life and maimings among the rural populace have already been alluded to. One must also point to the psychological toll of having to live with the constant danger of explosive remnants. Then there is the loss of access within uncleared areas to sources of water—so important in this part of the world. Indeed, some 450 wells and cisterns were for this reason inaccessible at the end of the war, a small number of them still so. Next must be mentioned the hampering of soil and water surveys and other rural planning activities and, of course, the actual cost of clearing the land of explosive remnants. And finally there are the losses, most difficult to quantify, which have been experienced by the regional wildlife.

IV. Conclusion

It has been shown that the explosive remnants of war can have an enormously adverse impact on an impoverished nation for many decades and thereby substantially impede its development. The remnants lead to losses that, first and foremost, can include thousands of direct human tragedies in the form of deaths and maimings. They can greatly exacerbate existing burdens of hunger and poverty. And they can do harm to wildlife. The explosive remnants of war are thus an inhumane and anti-ecological legacy of war that must be eliminated for the good of all.

References

Cestac, R. 1981. *Technical aspects related to material remnants of war in Libyan battle-fields.* Geneva: UN Inst. for Training & Research Publ. No. UNITAR/EUR/81/WR/3, 35 pp.

Gintzberger, G. & Bayoumi, M. 1972. *Survey of the present situation and production of the Libyan rangeland.* Tripoli: Agricultural Research Centre, 50 pp.

Le Houérou, H. N. 1965. *Improvement of natural pastures and fodder resources: report to the government of Libya.* Rome: Food and Agriculture Organization Expanded Programme of Technical Assistance Rept No. 1979, 46 pp.

Le Houérou, H. N. & Hoste, C. H. 1977. Rangeland production and annual rainfall relations in the Mediterranean basin and in the African Sahelo-sudanian zone. *Journal of Range Management,* Denver, **30**: 181–189.

Libya. 1981. *White book: some examples of the damages caused by the belligerents of the World War II to the people of the Jamahiriya.* Tripoli: Libyan Studies Centre, 176 pp.

Mostofi, A. 1983. *Remnants of war.* New York: UN Inst. for Training & Research Publ. No. UNITAR/CR/26, 124 pp.

Robb, R. L. 1945. *Survey of land resources in Tripolitania.* Tripoli: Stabilimento Poligrafico Maggi.

4. Explosive remnants of the Second Indochina War in Viet Nam and Laos

Earl S. Martin and Murray Hiebert
Mennonite Central Committee, Akron, Pennsylvania

I. Introduction

Huge amounts of high-explosive munitions were expended during the Second Indochina War, mostly by the USA. Thus, between 1965 and 1973 the USA expended a combined total of at least 14.3 million tonnes of air, ground and sea munitions in South Viet Nam, North Viet Nam, Kampuchea and Laos (table 4.1). This amount was almost twice the amount the USA had expended in all theatres during World War II, and is far in excess of munition expenditures by any single nation during any past or subsequent war. And, as in other wars, a significant fraction of the high-explosive munitions failed to explode, or were designed not to explode, on the occasion of their initial use.

The human costs of unexploded munitions from the Second Indochina War have been especially grave not only because of the enormous quantity of ordnance expended, but also because of the types of munitions used, the character of the war, and the nature of the theatre in which it was fought. A diverse array of munitions was employed, and some of the munitions were especially prone to malfunction. Few areas of Indochina remained exempt from both ground and air combat and hence from the subsequent scourge of unexploded ordnance.

This chapter focuses on the explosive remnants of southern Viet Nam and Laos, which together were the recipients of almost 90 per cent of the US munitions expended during the Second Indochina War (table 4.1). This is not to belie the fact that serious remnant problems were also created elsewhere in Indochina.

II. Southern Viet Nam

The USA used an estimated 10 million bombs and 220 million artillery shells

Table 4.1.   Munition expenditures by the USA in Indochina during 1965–1973 (in million tonnes = 10^9 kilograms)

Region	Air	Ground	Sea	Total
South Viet Nam	3.3	6.9	0.0	10.2
North Viet Nam	1.0	0.0	0.2	1.2
Kampuchea	0.6	0.1	0	0.7
Laos	2.2	0.0	0	2.2
Indochina	**7.1**	**7.0**	**0.2**	**14.3**

Source: Westing (1976, p. 14).

in South Viet Nam during the Second Indochina War (Westing, 1980, p. 101), along with over 100 million 40 millimetre grenades plus a vast array of other explosive anti-personnel munitions (Lindner, 1976; Lumsden, 1978; Prokosch, 1972; 1976).

The dud rate of munitions varies widely depending upon the type of ordnance and upon the conditions of use. Under field conditions—after munitions are adversely affected during shipping, storage and use by such factors as moisture, chemical corrosion, and quirks of the terrain and its vegetative cover—the average dud rate is often accepted to be approximately 10 per cent, and in some cases as high as 30 per cent or even more (Martin, 1973; Swearington, 1969). Using a value of 10 per cent, at least one million bombs, 22 million artillery shells, 10 million 40 millimetre grenades, and many millions of other anti-personnel bomblets remained unexploded at the end of the Second Indochina War as a horrible addition to the environment of South Viet Nam. Ordnance experts insist that each of these unexploded munitions must be considered armed and capable of detonation at any time in the future.

The human cost of these unexploded devices falls mainly on the rural population. Most typically, casualties occur when a farmer tills the soil and an unexploded remnant is stepped upon or is struck with a hoe or plough. Often children become casualties when they play with the curious metal objects they find (Dien Tin, 1974b; Martin, 1978, p. 180). To provide one actual incident, a 13-year-old boy, his brother and two young friends walked along a rice paddy dike on their way home from school one afternoon in 1974 in their village in northern South Viet Nam. The boy spotted a metallic object by the side of the dike the size and shape of a jumbo egg. Curious about the mysterious find, the group decided to take the object apart. When the boy struck it against a rock, the dud 40 millimetre grenade exploded. He was killed instantly. His brother escaped injury. The other two boys were rushed to the provincial hospital, but died within several days.

Fragmentary local statistics from South Viet Nam suggest that there have been thousands of victims of the explosive remnants of war, and that such casualties will continue into the future. One of the present authors (ESM) surveyed two districts of Quang Tri province (now part of Binh Tri Thien province) in 1974, when displaced farmers had been moving back for several months in order to reclaim abandoned fields after months of fighting in the region. Police records, believed by local people to be incomplete, listed 28 persons killed and 127 injured within the province during the previous six months. In the months following the end of the war in 1975, casualties from unexploded munitions were especially high among the people resettling the rural areas in this region. One report suggested 30 casualties per day in the province during this immediate post-war period. During the first post-war year a total of 886 thousand explosive remnants were neutralized in Quang Tri province. In 1983 a Vietnamese official reported that in Binh Tri Thien province (which straddles the former Demilitarized Zone between North and South Viet Nam and which has a population of 1.8 million) there had been some 5 000 casualties from explosive remnants since 1975 (Whitney, 1983).

The human toll from dud munitions and unexploded mines is compounded by the fear of movement in rural areas and by the denial of agricultural production and forest exploitation (Dien Tin, 1974a; Martin, 1978, p. 177). In some cases during the war the poorest members of society bore a dispropor-tionate share of the burden. Unable to pay rent for land, some farmers were permitted to plant rice rent-free in fields known to contain unexploded munitions. Moreover, civilians were drafted during the war by the authorities of the Republic of Viet Nam (the 'Saigon' regime) to clear heavily mined forest areas without proper training or equipment, resulting in deaths and serious injuries (Emerson, 1971).

Nature and distribution of explosive remnants

The types of explosive remnant varied from region to region in Indochina. Further, because of the nature of the war, it was not unusual to find one field free of explosives and an adjacent field littered with deadly debris. The explosive remnants in regions subjected only or primarily to aerial attack (North Viet Nam, large portions of Laos and Kampuchea, and small portions of South Viet Nam) consisted of large bombs and rockets, air-dropped mines, grenades and small anti-personnel bombs (bomblets). In regions where ground-fighting occurred (in many parts of South Viet Nam and in portions of eastern Kampuchea and south-eastern Laos), the problem of unexploded ordnance became more complex and unpredictable.

On the basis of a survey by one of the present authors (ESM) in Quang Ngai province (in what is now Nghia Binh province) in 1975 it became apparent that the distributional pattern of the explosive remnants basically consisted of five categories:

1. *Within the perimeter of former US or allied military outputs.* Here were found many mines planted to defend the outpost from attack. Although sometimes there would be anti-tank mines present, these were for the most

part anti-personnel mines. Most of the latter were the small 90 gram blast mine (model M-14) locally referred to as the 'toad' mine, but some were a larger 3.6 kilogram fragmentation mine (Model M-16). When set off by an individual, the former is likely to blow off a foot or do similar injury whereas the latter is likely to kill the person. In principle, when such a minefield is planted, the military unit doing so is expected to prepare a map indicating the position of the mines. In practice, those maps—if ever made—were not available in succeeding years, when farmers went back to reclaim their fields.

2. *Surrounding the perimeter of former US or allied military posts.* In the fields surrounding such outposts were found unexploded mortar shells of both sides as well as occasional booby traps. However, the most prevalent—and sub-sequently most dangerous—munitions found around these outposts were unexploded rounds of US 40 millimetre grenades, singled out for discussion below.

3. *Areas under the uncontested control of the former Republic of Viet Nam (the 'Saigon' regime).* The incidence of unexploded ordnance was generally not high in such areas.

4. *Areas under the uncontested control of the Provisional Revolutionary Government of the former Republic of South Viet Nam (the 'Viet Cong').* Some such areas were relatively free of unexploded munitions. Yet in many of these areas were found a large number of unexploded artillery shells, bombs, bomblets from cluster bomb units and air-dropped anti-personnel mines. In certain locations the revolutionary forces had also set land mines and booby traps for the defence of an area. The location of such devices were known to the local guerrillas and sympathetic civilians, but if such persons were killed or fled the area, the location of the explosives would not be known to subsequent residents.

5. *Contested areas.* Much of the rural region surveyed was more or less under the control of the 'Saigon' regime during the day, but during the night was essentially under the control of the revolutionary forces. Additionally, some small districts changed hands entirely from time to time. The problem of explosive remnants in such areas of disputed or changing control was most serious for at least three reasons: (*a*) the types of mines and other unexploded ordnance were diverse and unpredictable, both sides setting a variety of anti-personnel and anti-tank mines, and especially the 'Saigon' side expending many types of shells, bombs and grenades; (*b*) the quantity of munitions expended was far greater in a contested than uncontested area; and (*c*) much of the military activity was of a hit-and-run nature so that many of the battle zones were ephemeral and their locations not recorded.

One of the complicating factors in dealing with the explosive remnants of this war was that unexploded shells and bombs in areas which at one time had been under the control of the revolutionary forces were frequently utilized by the guerrillas to fashion new forms of explosive ordnance. For example, the guerrilla forces sometimes inserted a new explosive fuse into a defective artillery round and thereby converted it either into a land mine or a booby trap. Sometimes dud shells or bombs were cut in half with a hack saw and the explosive filler extracted for the manufacture of new land mines or booby traps (Martin, 1978, p. 182; Swearington, 1969).

The 40 millimetre grenade: a special case

Special attention must be accorded the US 40 millimetre grenade (e.g., model M-406) because it led to such a high number of post-war civilian casualties. During the war these high-explosive grenades were sometimes fired from helicopters, but most frequently they were fired by infantrymen using a hand-held grenade launcher (model M-79) having a range of 400 metres. These grenades, which look somewhat like a large egg, weigh about 225 grams and contain a high-explosive core wrapped with notched steel wire which breaks into numerous 130 milligram fragments when the grenades explode; the lethal radius is about 5 metres (Lumsden, 1978, pp. 132–133; Prokosch, 1972, pp. 19–23). An extremely large number of 40 millimetre grenades were fired during the war, apparently more than 140 million (Lindner, 1976).

Personal interviews and examinations of hospital records in former Quang Ngai province during 1974–1975 by one of the present authors (ESM) suggest that 40 millimetre grenades were the single most culpable explosive remnant in producing post-war civilian casualties. For example, during the two years following the cease-fire in 1973 eight persons were killed and 18 injured in an area of approximately 20 hectares surrounding a former US artillery position in Binh Son district (one of a dozen such outposts in the province). The local farmers attributed most of these 26 casualties to 40 millimetre grenade duds.

One farmer reported that he had picked up 50 unexploded 40 millimetre grenades as he tilled fields in the proximity of the Binh Son artillery outpost using a tractor equipped with a rotary tiller. Another farmer, who could not afford machine tilling, planned to till a 450 square metre field in the traditional manner of forcefully swinging a heavy broad-bladed hoe from above the head into the soil with the object of turning it. Before beginning he combed through the vegetation with his fingers and found 12 unexploded 40 millimetre grenades. The farmer carefully dislodged these rounds and disposed of them in an abandoned well. Nevertheless, when the farmer began to turn the soil his hoe struck a hidden grenade. The impact of the explosion threw him on his back and propelled the hoe through the air. Fortunately, the blade of the hoe and the earth had provided sufficient shielding so that the farmer emerged unscathed. As noted above, a number of his neighbours fared less well.

III. Laos

The USA dropped an estimated 6 million to 7 million bombs, plus huge but unknown numbers of bomblets on Laos during the Second Indochina War (Westing, 1980, p. 101). Perhaps 15 to 30 per cent of these were directed against northern Laos and the large remainder against south-eastern Laos, the so-called panhandle of Laos (IRC, 1975, p. 14265; Littauer & Uphoff, 1972, p. 281). To these figures must be added some relatively small though unknown amounts of ground munitions which were expended in Laos. One of the reported goals of the bombing in northern Laos was to destroy the social and economic infrastructure of the areas under the control of the anti-US Pathet Lao forces (Haan & Tinker, 1970, p. 19). A goal of the bombing in the panhandle was to interdict the so-called Ho Chi Minh trail being used as a supply route for the anti-US forces in South Viet Nam.

A region of northern Laos that was mercilessly bombed for years was Xieng Khouang province, especially its 140 000 hectare plateau often referred to as the Plain of Jars (Branfman, 1972; Hiebert & Hiebert, 1978; Swartzendruber, 1980). During the first five years after the bombing ended there in 1973, explosive remnants killed 267 persons and injured 343 others, according to government statistics provided on-site to one of the present authors (MH). For example, a 33-year-old rice farmer and father of two was working in his field in March 1979 near the Plain of Jars. He was hoeing the soil in preparation for the coming monsoon rains when suddenly his hoe struck a bomblet. The ensuing blast tore off his left hand. In 1980 alone, 46 people were reported to have been thus injured in the province and 23 killed. Similar figures have not been gathered for the whole of Laos. However, according to Laotian officials, the worst problems with explosive remnants have been faced by three provinces: the Xieng Khouang province just referred to; Houa Phan province, neighbouring it to the north-east; and Savannakhet province in the panhandle, especially its Muang Phin district. During the first half of 1979, 7 persons were reported to have been killed by explosive remnants in Houa Phan province and 14 in Savannakhet province.

The problem of explosive remnants not only threatens people's lives, but also hampers food production and post-war recovery in general. For example, Xieng Khouang province is about two million hectares in size of which about 300 000 hectares had been cultivated before the war. Almost one-third of this area had to remain out of cultivation as late as February 1981 owing to live ordnance, according to province officials.

Nature and distribution of explosive remnants

People returning to Xieng Khouang province and other war-ravaged areas after 1973 found a great number and variety of unexploded munitions. But not all of these explosives created equally serious hazards. Big bombs, artillery shells, and rocket and mortar rounds were for the most part relatively easy to detect and thus to avoid, defuse or detonate. As in southern Viet Nam, it was the numerous small anti-personnel weapons that have caused by far the largest numbers of casualties.

The most commonly encountered munition in Xieng Khouang province has been the cluster bomb (more properly, bomblet), which is colloquially referred to as the 'guava' bomb (e.g., model BLU-26) and what the Laotians have commonly referred to as the 'bombie' (Lumsden, 1978, pp. 145–161; Prokosch, 1972, pp. 43–47; 1976). These are packaged for air drop in groups of perhaps 670 in a bomb-shaped dispenser (e.g., model SUU-30), the combined package (weighing about 380 kilograms) being referred to as a cluster bomb unit (e.g., model CBU-24).

The soft-metal casing of the 'guava' bomb has embedded in it about 300 6 millimetre diameter steel balls which fly out in all directions when the bomblet explodes, with a lethal radius of over 5 metres. The dispenser is designed to split open before it reaches the ground so that the bomblets become distributed over an area of perhaps 0.5 hectare or more. The bomblet can be equipped with one of several types of fuses.

The second most commonly encountered munition in Xieng Khouang province has been the so-called 'pineapple' bomb which is packaged for air drop in groups of 360 (the bomblet being, e.g., model BLU-3; the dispenser, e.g., model SUU-7; and the combined unit, e.g., model CBU-2). These are cylindrical bomblets with stabilizing fins whose soft-metal walls each contain about 250 small steel balls (Lumsden, 1978, pp. 145–161; Prokosch, 1972, pp. 43–47).

Several other types of small fragmentation munition, variously fused, were also commonly dropped on Laos as cluster bomb units (Krepon, 1973–1974; Lumsden, 1978, pp. 145–161; Prokosch, 1972, pp. 43–52). The colloquially named 'orange' bomb (model BLU-24), weighing 730 grams, was designed to penetrate a jungle canopy and reach the ground before exploding. Here fragments of the casing, rather than embedded pellets, cause the damage. The so-called 'butterfly' bomb, weighing about 1.8 kilograms, was also dispensed in a cluster bomb unit. Another bomblet that was used (model BLU-61) is similar in appearance to a 'guava' bomb but somewhat larger, weighs 1.0 kilogram, and is a combination fragmentation/incendiary device.

Also dropped in huge numbers for purposes of area denial were tiny anti-personnel mines referred to as 'dragontooth' mines (models BLU-43 and BLU-44). These minelets are blast weapons weighing only about 20 grams each, but when stepped on are capable of tearing off the foot. A cluster bomb unit contains about 4 800 dragontooth mines. Also much used was the 'spider' mine (model BLU-42) which looks similar to the 'guava' bomb, but sends out eight trip wires after being dropped and is not meant to go off until these are disturbed.

IV.   Measures to eliminate the unexploded munitions

Eliminating unexploded munitions under the rural post-war conditions of Indochina has proved to be highly difficult. Detection—the first step in the process—is dangerous and tedious. Metal detectors, which are reasonably useful in locating emplaced land mines, have not been available to the farmers. At any rate, metal detectors are of little use for locating non-metallic mines and in areas littered with anti-personnel bomblets owing to the inevitable presence of numerous metal fragments from previously exploded munitions. The use of pointed metal rods for probing the undergrowth and earth in search of explosive remnants requires special training. Farmers thus have had to rely on visual means of detection, sometimes where feasible made easier by first burning off the vegetation. Once detected, explosive ordnance has been marked and avoided or else carefully carried to a remote location for burial (or ultimate neutralization by experts).

As suggested earlier, most of the large bombs are relatively easy to detect and are then avoided until they are neutralized, usually by military personnel. It is the small anti-personnel munitions that have caused so much grief. These are generally hidden in the undergrowth or buried shallowly. Accidents occur when a farmer is hoeing a field, when workers are preparing the ground for road or house construction, when children play with objects they find, and occasionally when an outdoor fire is lit.

Little in the way of systematic techniques has been developed in Indochina for dealing with small unexploded ordnance. In Laos in 1977 one of the present authors (MH) observed a group of secondary-school teachers and students prepare the ground for rebuilding their school in Muang Kham district of Xieng Khouang province. As a bulldozer cleared the brush, people followed to pick up carefully the exposed bomblets and gingerly carry them to a leaf-lined hole for disposal. Similarly, when villagers elsewhere in the province were preparing to rebuild, they threw bomblets they had uncovered into an open pit. Sometimes the bomblets would explode on impact; otherwise they were subsequently detonated using a fire.

US military sources strenuously warn against the casual handling of anti-personnel bomblets. Dud bomblets which malfunctioned at the time of initial use and have subsequently corroded in the field are particularly unpredictable and thus mortally dangerous. They can be sensitive to pressure, to rotation or other movement, and to small shocks. If at all possible, the bomblets should be detonated in place by remote means. They should be physically handled only by experts and only as a last resort. For certain ones (for example, the 'pineapple' bomb) it is recommended that, if moving is required, they first be encased in plaster of Paris in order to prevent the fuse from becoming activated.

The safest method of disposing of an explosive remnant is to blow it up in place, usually with a 100 gram charge of TNT. However, this requires: (*a*) someone who is trained to do so; and (*b*) the necessary explosive and associated paraphernalia. These conditions have typically not as yet been met in South Viet Nam or Laos.

A second method of disposal is to disarm the munition by removing the activating device or devices. This is an extremely hazardous operation because of the unknown nature of the original malfunction, because of subsequent corrosion, and because of internal fusing, anti-disturbance fusing or other booby-trapping, and so forth. This approach is therefore limited to very special circumstances and highly-trained personnel; it has no general applicability.

An approach that is very useful, where the terrain conditions and availability of equipment permit, is to plough fields using heavy well-shielded equipment such as armoured tractors, personnel carriers or tanks. However, this method of rendering a field more or less safe for subsequent hand tilling has been possible on only a limited basis.

The US Department of Defense suggested to the authors (in private correspondence dated 14 March 1978) that one sequential procedure to follow in clearing an area of explosive remnants is to: (*a*) burn off the ground cover; (*b*) make a visual search, detonating in place any munitions discovered; (*c*) sweep with an electronic metal detector, verifying any contacts made by careful excavation, and detonating in place; and (*d*) plough or harrow using a tractor (or perhaps draught animal), again detonating in place any further munitions discovered.

Unfortunately the procedure just outlined is quite impractical in rural Indochina for the simple reason that the explosives needed to detonate the munitions in place, the electronic metal detectors needed to discover the buried ordnance, the tractors needed to help with the final search, and the trained disposal personnel needed to carry out all of this are usually not available.

In Laos in 1979 the USSR initiated an aid programme to clear farm land in Xieng Khouang province. The programme involved 12 Soviet experts plus 120 Laotian trainees for a period of 18 months. The major approach was to employ metal detectors mounted on the front of jeeps which were driven slowly over the area to be cleared. When a signal was received the spot was marked and the munition subsequently detonated in place by a separate team. During the 18 months some 5 000 hectares were cleared of 12 700 explosive remnants of many types (with the 'guava' bomb predominating). No casualty was incurred during the operation. The Laotians, who had been trained, subsequently spread out over five provinces, but were severely hampered in their work because they ran out of explosives and because of a lack of batteries to run the five metal detectors they had received.

Two non-governmental US agencies—the American Friends Service Committee (Philadelphia, Pennsylvania) and the Mennonite Central Committee (Akron, Pennsylvania)—have provided some relevant assistance in both Viet Nam and Laos. For example, in the latter country wide-pronged garden forks and, later, spades were provided to farmers in Xieng Khouang province for the purpose of tilling the land more gently than with the traditional hoes. Subsequently a tractor was also provided that had been equipped with a specially designed flail mounted in front (a rotating axle to which was welded short lengths of heavy chain), an automatic brush-cutter mounted on the back (a 'rotovator'), and steel plates mounted so as to protect the cab (Swartzendruber, 1980). Unfortunately, the tractor did not prove to be very successful in setting off the anti-personnel munitions it encountered, despite a number of on-site design modifications. The tractor project was in time abandoned owing to a combination of technical and bureaucratic problems.

V. Conclusion

The Second Indochina War has highlighted the difficulties faced by civilians who try to deal with the problem of explosive remnants of war with only minimal information about the types of munition involved and their wartime distribution and with no sophisticated equipment for dealing with them. What has also been demonstrated is the difficulty of coping with a highly technical problem in a non-technical, underdeveloped society in which the authorities face many other problems considered to be of equal or higher priority.

It became clear that the problem of unexploded munitions is really a problem too large and complex to be addressed adequately by private agencies such as the Mennonite Central Committee. Ideally, it is the military forces which use the munitions in a war that should accept responsibility for ultimately disposing of the unexploded remnants.[1] However, inasmuch as military forces usually refuse to do so for various political or military reasons, ordnance specialists from neutral countries or from an international body such as the United Nations should be available to give appropriate advice and assistance.

Alternatively, or additionally, training in munition disposal should be made available to personnel in private agencies that have demonstrated a willingness to assist in the clearing of explosive remnants of war, something which the USA refused to do for the Mennonite Central Committee at the end of the Second Indochina War.

Moreover, it is essential that at the termination of hostilities military forces provide the affected countries with at least elementary information about the nature of the ordnance used and with safe methods for its disposal.

Unexploded munitions will cause civilian casualties wherever war occurs. But one of the lessons of the Second Indochina War has been that certain munitions—for example, the 40 millimetre grenade and the 'guava' bomb—are especially pernicious in that regard. If nations must continue to resort to war to address their international problems, then such especially inhumane weapons should be outlawed.

---

[1]A step in the direction of accepting post-war responsibility for clearing the explosive remnants of war left behind was, in fact, taken at the termination of the Second Indochina War. According to one of the protocols that accompanied the 1973 agreement ending that war, the USA accepted the responsibility to clear Haiphong harbour and other territorial waters of North Viet Nam of the mines it had dropped there (see appendix 7, sect. II), and subsequently carried out this obligation. The USA accepted the further responsibility to clear the land of South Viet Nam of its mines and other explosive remnants (see appendix 7, sect. III), but this obligation was for the most part not carried out.

References

Branfman, F. (ed.). 1972. *Voices from the Plain of Jars: life under an air war.* New York: Harper & Row, 160 pp.

Dien Tin. 1974a. [Clear mines to return 300 hectares of land to people for use.] (In Vietnamese). *Dien Tin*, Saigon, **1974** (16 Jul).

Dien Tin. 1974b. [Five children play with M79 grenade leaving one dead four wounded.] (In Vietnamese). *Dien Tin*, Saigon, **1974** (4 Jun).

Emerson, G. 1971. Villagers say Saigon perils their lives. *New York Times* **1971** (10 Jan): 1, 3.

Haan, D. S. de & Tinker, J. M. 1970. *Refugee and civilian war casualty problems in Indochina: a staff report.* Washington: US Senate Committee on the Judiciary, 107 pp.

Hiebert, L. & Hiebert, M. 1978. Laos recovers from America's war. *Southeast Asia Chronicle*, Berkeley, California, **1978**(61): 1–20.

IRC (Indochina Resource Center). 1975. Indochina war statistics: dollars and deaths. *US Congressional Record*, Washington, **121**: 14262–14266.

Krepon, M. 1973–1974. Weapons potentially inhumane: the case of cluster bombs. *Foreign Affairs*, New York, **52**: 595–611.

Lindner, V. 1976. Engineering miracle in munition design. *National Defense*, Washington, **60**: 294–297.

Littauer, R. & Uphoff, N. (eds). 1972. *Air war in Indochina.* rev. ed. Boston: Beacon Press, 289 pp.

Lumsden, M. 1978. *Anti-personnel weapons.* London: Taylor & Francis, 299 pp. [a SIPRI book].

Martin, E. 1973. Defusing the rice paddies. *Washington Post* **1973** (8 Jul): D6.

Martin, E. S. 1978. *Reaching the other side: the journal of an American who stayed to witness Vietnam's postwar transition.* New York: Crown, 282 pp.

Prokosch, E. 1972. *Simple art of murder: antipersonnel weapons and their developers.* Philadelphia: American Friends Service Committee, 88 pp.

Prokosch, E. 1976. Antipersonnel weapons. *International Social Science Journal*, Paris, **28**: 341–358.

Swartzendruber, J. 1980. Tractors in Laos: picking up the pieces. *Christianity and Crisis*, New York, **40**: 243–246.

Swearington, [T.] 1969. *Staff study on pernicious characteristics of U.S. explosive ordnance.* Washington: US Marine Corps, unpubl. ms (Oct 1969), 10 pp.

Westing, A. H. 1976. *Ecological consequences of the Second Indochina War.* Stockholm: Almqvist & Wiksell, 119 pp. + 8 pl. [a SIPRI book].

Westing, A. H. 1980. *Warfare in a fragile world: military impact on the human environment.* London: Taylor & Francis, 249 pp. [a SIPRI book].

Whitney, C. R. 1983. Bitter peace: life in Vietnam. *New York Times Magazine* **1983** (30 Oct): 24–27, 54–63.

5.   Explosive remnants of war on land: technical aspects of disposal

Bengt Anderberg
Swedish International Humanitarian Law Delegation

I.   Introduction

Large amounts of explosive ordnance, both mines and unexploded bombs and shells, are left behind following a war. In many countries the environmental impact of these remnants is both widespread and of long duration. Although many of the afflicted nations have adopted comprehensive countermeasures and have been more or less successful in clearing such ordnance away, others do not have the technical capacity to do so.

The technical problems of clearing explosive remnants of war are stupendous. Clearance remains time-consuming, costly, and to some considerable extent dangerous. As a result of technical advances in mine development, those mines that are left behind have become ever harder to discover and clear. At the same time it is becoming easier to mine a large land area rapidly.

This chapter begins by describing both the mines and other munitions that become the explosive remnants of land warfare (the remnants of naval warfare being covered in chapter 6). Next described are the available means of disposal. The chapter concludes with some proposals for alleviating the problem. The present work draws to some extent on a survey conducted by the author on behalf of UNEP (Tolba, 1977) and on other of his previous studies (Anderberg, 1981a; 1981b).

II.   Explosive remnants of land war

Developments in military strategy and technology over the last hundred years have resulted in a steadily increasing use of ever more complicated explosive munitions. Both the undetonated mines and the bombs and shells which did not explode at the time of use can remain dangerous for very long periods of time. Moreover, the increasing technical complexity of explosive munitions has led to an ever growing proportion of such dangerous duds.

This section provides a survey of manually emplaced land mines, both anti-tank and anti-personnel, of remotely delivered anti-tank and anti-personnel mines, of booby traps, and of unexploded munitions.

Emplaced anti-tank mines

When during World War I tanks began to be used in ever increasing numbers, there arose a need for mines that could halt a tank attack. The first anti-tank mines, developed in the latter part of the war, consisted of simple containers filled with explosives whose blast effect was sufficient to destroy the track of a tank as long as it exploded directly underneath it.

Very soon mines and their fuses were constructed of a size and form that guaranteed their function even if the tank track did not run over the whole mine. Rather sophisticated models were available by the time World War II began (Tresckow, 1975). An important example was the German anti-tank mine model T-Mi35. This mine had a sheet-metal casing and weighed 9 kilograms of which 5 kilograms was the explosive filler; it was 10 centimetres high and 32 centimetres in diameter. It was equipped with a pressure-sensitive lid which caused the mine to detonate when subjected to a load of at least 100–190 kilograms.

Different variants of this anti-tank mine were used in very large quantities by a number of the belligerents during World War II. In addition to these standard anti-tank mines, others were manufactured which contained very little metal or even no metal at all (Tresckow, 1975). For example, there was an Italian mine made with a bakelite plastic casing, a French mine with a glass casing, and various other French mines with wooden casings. The type of casing has a strong bearing on the longevity of a mine as well as on its detectability by metallic mine detectors.

In the development of anti-tank mines following World War II many countries concentrated on making them non-metallic to the extent possible. Most such mines do contain some metal, especially in the fuse. However, a number of years after the Korean War the USA introduced its completely non-metallic model M-19 for which even a plastic spring for the fuse had been developed (Tresckow, 1975).

The generation of anti-tank mine just described, whether metallic or non-metallic, still depends upon the direct pressure of a tank track for detonation. In consequence, an enormous number of such mines have to be emplaced in order to obtain an adequate level of obstruction. But there is an even more significant consequence: such a mine generally damages the track alone, leaving the hull, engine and transmission essentially undamaged and the crew often unhurt. Thus, during the Fourth Arab-Israeli ('Yom Kippur') War of 1973 approximately 75 per cent of the Israeli tanks which had been disabled by pressure mines were operational again 24 hours later (Crèvecoeur, 1982).

These disadvantages led to the development of a new generation of anti-tank mine with more sophisticated sensors and having an impact on the whole breadth of the tank (Crèvecoeur, 1982; Golino, 1981). One or more sensors register seismic, magnetic or other disturbance. If the approach of a valid target is indicated, the mine detonates. The explosive element is a vertically directed so-called hollow charge which does not merely disable a tank, but destroys it. It is very important to point out that these new mines are in some instances being manufactured to have an operational life of between 30 and 180 days. For example, the French model HPD1A ceases to function after 60 days. Alternatively, the Italian model TCE/6 can be made operative by a radio signal and also rendered inert in the same way (Pengelley, 1982).

A number of special-purpose anti-tank mines have also been developed. For example, the Federal Republic of Germany has developed one for emplacement in rivers for use against wading or floating tanks. Another special-purpose mine is the French model MAHF1 which propels a warhead horizontally up to 40 metres to strike an oncoming tank.

Emplaced anti-personnel mines

Anti-personnel mines are usually fitted with a charge of about 150 grams, an amount quite sufficient to disable or kill a person. Even during World War II many different variants existed. Some mines (fragmentation mines) projected either a shower of metal fragments or shot (pellets) when their trip wire was disturbed. An especially fearful variant was the bounding mine which was thrown up some 1.5 metres before detonating. Its fragments had a lethal radius of up to 10 metres and could cause severe injury even at a distance of 50 metres. The prototype for the bounding mine was the German model S-Mi35, subsequently copied by many other countries. Other mines (blast mines) were set off when an individual stepped on them, the device then blowing the person's foot to pieces. Whereas the fragmentation anti-personnel mines were metallic, the blast anti-personnel mines were generally non-metallic (e.g., the Soviet model PMD-6 with a wooden casing or its model PMK-40 with a cardboard casing). Some anti-personnel mines of either type (although especially of the blast type) were of such high-quality manufacture that they are probably still in working order today in the World War II minefields that remain, especially in desert areas.

Anti-personnel mines have been developed along different lines since World War II (Tresckow, 1975). The pressure-fused mines have been constructed to contain very little or even absolutely no metal. The US model AP-M14 is made of plastic with only the tip of the striker being of metal. It can be armed and disarmed several times. A number of these mines have been given such a simple construction that it is not possible to render them inert once they have been armed, for example the Italian model SACI56. Some modern fragmentation mines give off a directed shower of steel shot having a great penetrative capability. The bounding fragmentation mine described above is still being manufactured in roughly the same design as in World War II. Some anti-personnel mines can be set off not only as noted earlier, but also by remote electric or radio signal when an observer chooses the right moment.

Remotely delivered anti-tank and anti-personnel mines

Technical developments in mine warfare which are of great importance regarding the explosive remnants of future wars are now taking place very rapidly. Both anti-matériel and anti-personnel mine systems that can be remotely emplaced in large numbers with mortars, artillery, rockets, missiles, aircraft and other dispensers are being developed on an extensive scale by various major nations.

Mines fully capable of penetrating the bottom armour of a tank can now be made compact enough for nine to fit into a 155 millimetre howitzer shell (McDavitt, 1979). Similarly, one such shell can take 36 trip-wire activated anti-personnel mines. The Federal Republic of Germany has a 110 millimetre rocket which can be launched from a 36-barrel launcher to a range of 14 kilometres. Each rocket can deliver eight anti-matériel (model AT1) or five anti-tank (model AT2) mines. It has a similar launcher which can be mounted on either a tracked carrier or helicopter, as do several other countries, for example, the USA and Italy (Pengelley, 1982). These examples will give an indication of present and future possibilities.

Most of the remotely delivered (scatterable) mines now being developed are designed to destroy themselves at a pre-determined time. This is necessary in part for operative reasons, but it is also a requirement of the 1981 Inhumane Weapons Convention (see appendix 3). In this regard it must be pointed out that some more or less small proportion of the mines which are meant to become harmless in time will fail to do so. Thus there will still be dangerous remnants even when these systems are employed.

Booby traps

Manually emplaced mines are emplaced so as to defy discovery by the enemy until such time as they go off as intended, that is, to destroy a tank or other vehicle or to kill or maim an individual. Moreover, anti-personnel mines are often emplaced among anti-tank mines in order to impede their neutralization.

In addition to the procedures just alluded to, mines are often fitted with devices—known as booby traps—that will set them off when they are tampered with. Such arrangements to prevent the clearance of minefields will always make emplaced mines dangerous either to move or to defuse in place.

Another type of booby trap that can be found in a theatre of war is an apparently innocent object which is fitted with an explosive device.

Unexploded munitions

Huge amounts of munitions are expended in warfare. Some fraction of these always fail to explode at the time of initial use and remains in the battlefield as potentially explosive remnants for many years after a war. Many countries today suffer severe deterioration of their environment from these duds.

The Netherlands has estimated that between 5 and 10 per cent of the total munitions expended in its country during World War II failed to explode when intended (unpublished letter from the Netherlands to UNEP dated 22 November 1976). The failure rate of US munitions under the wet and otherwise adverse weather conditions that existed during the Second Indochina War was perhaps three times worse, especially for munitions fitted with complex fuses (Swearington, 1969).

III. Disposal of explosive remnants on land

The technical problems involved in the systematic disposal of the explosive remnants in a former theatre of war are truly formidable. A huge amount of remarkably different devices can be dispersed over wide areas. Not all material remnants are dangerous, but before an object can be considered harmless, it has to be investigated. This means that each remnant must be located, investigated, identified, and—if a danger—neutralized.

Many governments maintain specially trained and equipped military units (usually a branch of military engineers) for ordnance disposal. Numerous clearance operations have had to be carried out in various parts of the world since World War II, and a number of currently in progress (see, e.g., chapter 2). There are thus a great number of experienced personnel scattered throughout various countries. A great deal of relevant information is in the public domain, but the details of some methods are kept secret.

The disposal of explosive remnants is here discussed under three headings: planning, detection and neutralization.

Planning

The disposal of explosive munitions calls for careful planning based on as much background information as possible. It is necessary that all relevant technical information be at hand and that all existing historical information be made available. Information on the geography of the area, its weather conditions, and its resident population is also of significance. The level of regional and national support and assistance to be expected must be ascertained as well.

Examples of some of the questions that arise in the planning stage of clearing a minefield are presented here just in order to give an indication of the complexities involved in any clearance operation:

1. Is there any documentation available (minefield establishment maps or reports, wartime diaries, accident records, etc.)?

2. What is the total number of mines emplaced; what types of mines are involved; are technical descriptions available; where is each individual mine situated; and are the mines booby-trapped?

3. Does the minefield also contain unexploded (dud) munitions; and, if so, of what types and in approximately what numbers; and, again, are technical descriptions available?

4. What has happened to the minefield since it was laid; have there been soil changes (soil loss from flooding, etc.; soil accretion from sandstorms, etc.); have there been changes in the vegetation; and so forth?

5. Can the area be readily reached by the disposal team; are people living in the danger area; are temporary quarters available; and are supplies locally available?

6. Which authorities are responsible for the clearance operation; who is to supply the disposal personnel, the necessary equipment, the logistical support and the financing; and within what time frame is the operation to be carried out?

Completely satisfactory answers to many of these questions—and to others not listed—will, of course, not be possible. This means that the planning of a clearance operation will always have to take place against a background of considerable uncertainty.

Detection

The most primitive method for the detection of explosive remnants is to divide the land to be searched into small squares, for example, 20×20 centimetres in size, and then to probe each of the squares with a sharpened stick or rod. This is a very time-consuming and somewhat dangerous method. Moreover, it is not possible to detect deeply buried explosive ordnance by this means.

Metallic mine detectors were developed during World War II and are used to locate metallic mines and duds (Ludvigsen, 1982, pp. 47–53). The principal hand-held metal detector of the US armed forces is model AN/PSS-12 (Foss & Gander, 1984, p. 236). Metal detectors cannot usually distinguish between an explosive munition and metal fragments, for example, from those of a munition that has already exploded. This lack of discrimination is a severe limitation since a profusion of metal fragments is likely to be prevalent wherever mines and duds exist.

With the introduction of non-metallic mines, more sophisticated techniques have had to be explored. There have been developed very sensitive metal detectors capable of locating plastic mines that contain only small amounts of metal. Other detectors depend upon sensing changes in the soil. One current version is the hand-held US model AN/PRS-7 (Foss & Gander,

1984, p. 235). Another is the vehicle-mounted US model AN/VRS-5 which is a microwave device that locates mines by detecting differences in reflectivity between mine and soil (Foss & Gander, 1984, p. 236; Hughes, 1979). However, it must be pointed out that the non-metallic detectors are difficult to operate and that they give falsely positive signals for such buried items as stones and large pieces of shrapnel. A very sophisticated detector (designated model MSG1) is said to be under development in the Federal Republic of Germany (Pengelley, 1982).

Detectors are being developed that can sense the vapours given off by the explosive charge within a munition. An example of such an electronic 'sniffer' is the British model SA23 (IDR, 1976, p. 821).

One of the best means for detecting explosive remnants is with the help of specially trained dogs under the control of comparably specially trained handlers (see chapter 7). The dog locates the ordnance through its ability to smell the explosive. It thus can find a non-metallic mine which the electronic detector overlooks and will ignore the shard of metal that the electronic detector senses. The use of dogs is faster and safer than other methods, although it is by no means foolproof. Few such dogs are available today because the training of a dog plus its handler is a lengthy and exacting affair and because under operational conditions the dog can work for only a brief period each day and during that time requires a relatively quiet, pressure-free and non-distracting environment. The use of other animals, for example, rats, is under consideration (Meyer, 1982b).

Mechanically laid minefields are difficult to conceal from the air when newly emplaced. Infra-red ('false colour') photography and other infra-red scanning devices readily detect disturbed earth and vegetation (Kitching, 1977). However, this approach has limited, if any, value in the detection of mines in old minefields, of remotely delivered mines, or of unexploded (dud) munitions.

Various possibilities are being considered in the development of new detectors. Among these are sensing mechanisms based on X or gamma radiation, on infra-red radiation, or on radar (Meyer, 1982a).

Neutralization

Once detected, the basic principle is to blow up the explosive munition in place. However, this is a slow operation when it comes to areas containing a large number of remnants. The armed forces of some countries have devices or methods for neutralizing small areas which are either operational or under development (Kitching, 1977). These systems are designed for wartime battle-field application and are not generally suitable for use in peace-time. A number are described below.

The mine-clearing roller is a set of three or four heavy steel discs mounted in front of each track of a combat vehicle. Each disc has freedom to exert pressure on the ground independently as it follows the terrain and also to recoil independently when a mine is detonated. A mine-clearing plough can be attached in tandem with the roller to deal with the few mines that escape the roller (Williams, 1979). The roller is meant to set off mines which are actuated by pressure, vibration or magnetic influence and the plough to cast aside those that are missed. The plough can also be used independently. The mine-clearing flail is another device mounted in front of a vehicle in which a series of swinging chains beats the ground. These devices are all designed to clear a narrow path through a minefield, but are ill suited for wide-area clearance.

A number of explosive devices are used by armed forces for the breaching of minefields. Some of them take the form of an explosive-filled hose or pipe or of a string of charges; they are rocketed into position and then set off. The blast effect clears a narrow path. The US model M58A1 is said to be capable of clearing a path 8 metres wide and 100 metres long (Pengelley, 1982). Alternatively, the fuel–air explosive devices disperse a highly volatile liquid as an aerosol cloud (having a diameter of, say, 12 metres) over a minefield. When detonated the blast is meant to set off the pressure-sensitive mines and booby traps beneath the cloud. In one recently developed system a 30-tube 345 millimetre rocket launcher sends its rockets in a linear pattern which is meant to clear a lane 8 metres wide and 240 metres long (Pengelley, 1982). In general, explosive methods, whether based on line charges or fuel–air explosives, are mainly suitable for neutralizing mines actuated by simple pressure fuses and cannot be relied upon to set off other mines or dud munitions. Moreover, they themselves add to the environmental disruption.

As already stated, the basic principle is to blow up a dud or mine on the spot. Sometimes this is not possible. It must then often be disarmed by defusing. This requires great skill since it is extremely dangerous to disturb an explosive with an old and perhaps faulty fuse. Moreover, a mine or even an unexploded bomb is often fitted with an anti-disturbance device or might be otherwise booby-trapped. Munition disposal personnel use a variety of protective, diagnostic, remote-viewing, remote-handling and other safety devices. It is important to reiterate that so far no technical means is known by which the systematic clearance of an area can be accomplished in a rapid, safe and simple way.

IV. Conclusion

The many wars of the twentieth century have resulted in enormous amounts of explosive remnants. These munitions have seriously limited the human environment. Large areas have been shut off for long periods. Agriculture, forestry, fishing, mining, and so forth have been prevented or made difficult. Many people have been killed or maimed, including a disproportionately high number of children. The aggregate human and environmental impacts have been immeasurably large.

Very large amounts of explosive munitions remain today in many countries despite continuing efforts by these countries to rid themselves of them. Many developing countries especially have insufficient economic resources, technical expertise, or personnel to deal adequately with the problem.

The increasing military emphasis on wide-area weapons which contain many sub-warheads is drastically increasing the number of explosive remnants, numbers that are further multiplied because their increasing complexity leads to a greater proportion of malfunctionings. It does appear as if the large numbers of remotely delivered (scatterable) mines will be equipped with self-inactivation mechanisms, but some fraction of these mechanisms will not function, perhaps again to produce large numbers of explosive remnants.

Given these circumstances, it is proposed that an international agency within the United Nations system (perhaps UNEP) should work towards achieving the following three aims: (*a*) clearance of the existing explosive remnants of war that constitute a threat to the human environment; (*b*) wider acceptance of the 1981 Inhumane Weapons Convention and a broadening of the scope of this Convention; and (*c*) a mechanism for the rapid clearance of explosive remnants resulting from future wars.

Clearance of existing and future explosive munitions calls for an international staff that consists of military, technical, legal, environmental and other experts. This body would collect information on explosive remnants of war and on the various material and personnel resources that are available for clearance. It ought to be possible for the body to take the initiative, in co-operation with the countries affected, to make proposals on a clearance procedure, a procedure which should be set in motion immediately following the cessation of active hostilities. In addition, the body must be able to keep up with technical developments in both weapon systems and clearance systems.

As to strengthening the 1981 Inhumane Weapons Convention, it is urged that the parties convene a conference, as provided for in Article 8 (see appendix 3), with an aim, *inter alia*, to lessen the danger of the explosive remnants of war. It is necessary to strengthen and expand Protocol II of the Convention, and the possibilities for accomplishing this appear good. The following items are suggested for consideration:

1. That it be made compulsory for mines (both land and sea) to be fitted with a device that in time automatically renders them harmless, or else with the means for receiving a remote signal that does so.

2. That it be made compulsory for the belligerents to keep a set of specified records of minefields laid (whether on land or sea) that include not only the location of the field, but also technical particulars of the mines emplaced; and that these records be made available at the cessation of hostilities.

3. That it be made compulsory for technical information to be made available by the belligerents at the cessation of hostilities on the explosive

munitions employed in order to facilitate the clearance of duds; and information also to be made available on likely areas of concentration and on recommended means of detection and disposal.

4. That it be made compulsory for the belligerents to provide information at the cessation of hostilities on abandoned munition caches, munition dumps (on both land and at sea), on shipwrecks containing munitions, and on any other obviously dangerous material remnants of the war.

References

Anderberg, B. 1981a. *Communication.* Geneva: UN Inst. for Training & Research Publ. No. UNITAR/EUR/81/WR/20, 5 pp.

Anderberg, B. 1981b. *Disposal of material remnants of wars.* Geneva: UN Inst. for Training & Research Publ. No. UNITAR/EUR/81/WR/22, 3 pp.

Crèvecoeur, P. 1982. Vertical anti-tank defence. *Armada International,* Zurich, **6**(3): 57, 60, 62.

Foss, C. F. & Gander, T. J. (eds). 1984. *Jane's military vehicles and ground support equipment.* 5th ed. London: Janes's Publishing Co., 871 pp.

Golino, L. 1981. New advances in mine warfare. *Defence Today,* Rome, **5**: 495–497.

Hughes, B. C. 1979. Combat engineering equipment for the 1980's. *Military Engineer,* Washington, **71**: 406–410.

IDR (International Defense Review). 1976. Internal security equipment. *International Defense Review,* Geneva, **9**: 820–823.

Kitching, J. 1977. Minefield breaching. *International Defense Review,* Geneva, **10**: 523–525.

Ludvigsen, E. C. 1982. 'Support forward': when and if the budget permits. *Army,* Arlington, Virginia, **32**(5): 34–37, 41–43, 47–49, 53.

McDavitt, P. W. 1979. Scatterable mines: superweapon? *National Defense,* Arlington, Virginia, **64**(356): 33–37.

Meyer, D. G. 1982a. Deadly game of 'hide and seek' using World War II technology? *Armed Forces Journal International,* Washington, **119**(7): 22, 24.

Meyer, D. G. 1982b. Sniffing out dangerous mines a real turn-on for rats. *Armed Forces Journal International,* Washington, **119**(7): 24.

Pengelley, R. 1982. Mining and counter-mining: paths to the future. *Defence Attaché,* London, **1982**(1): 9–20.

Swearington, [T.] 1969. *Staff study on pernicious characteristics of U.S. explosive ordnance.* Washington: US Marine Corps, unpubl. ms (Oct 1969), 10 pp.

Tolba, M. K. 1977. *Implementation of General Assembly resolution 3435 (XXX): study of the problem of the material remnants of war, particularly mines, and their effect on the environment.* Nairobi: UN Environment Programme Document No. UNEP/GC/103 (19 Apr 1977), 8 pp. + UNEP/GC/103/Corr. 1 (6 May 1977), 1 p. Also: New York: UN General Assembly Document No. A/32/137 (27 Jul 1977), 1 + 8 pp.

Treskow, A.v. 1975. [Land mines.] (In German). *Soldat und Technik,* Frankfurt a.M., **18**: 388–400.

Williams, R. N. 1979. Finding and clearing mines. *Armor,* Washington, **88**(6): 12–16.

6.   Explosive remnants of war at sea: technical aspects of disposal

Willard F. Searle, Jr and Dewitt H. Moody
United States Navy (Retired)

I.   Introduction

This chapter presents a brief discussion on technology to be used by qualified personnel for the clearance of the explosive remnants of war from inland waterways, coastal waters and oceans. Although the subject is addressed here in general terms, it must be emphasized that each event will be different so that the techniques and equipment employed must be tailored to the operational environment, the remnants located and the level of clearance required.

The chapter first notes problems typical of various parts of the world. This is followed by background information that should be considered in the planning of clearance operations. Additionally discussed are the needed technical training and the required technology for locating and removing or disposing of the hazardous items in question. Past work on clearance has been focused largely on clearance of land rather than the aquatic environment (see appendix 1). This chapter is meant to help rectify that omission. It is based to some extent on a prior consideration of the subject by the present authors (Searle & Moody, 1981).

II.   The problem

Various offshore areas are designated on standard hydrographic charts and piloting instructions as danger areas because of old minefields or other hazardous remnants of war such as sunken barges or ships and explosive or chemical ordnance. Such areas are closed generally to navigation, fishing and work on the seafloor. By way of example, "There are two areas declared dangerous due to mines in Gulf of Suez: (i) Close N of Râs Abu Bakr (28°33'N, 32°56'E) . . . (ii) On the NE side of Strait of Gûbal . . . Other mined areas may exist in Gulf of Suez and in Suez Bay; information should be sought from local authorities before arrival" (Aldridge, 1980, p. 1). Similarly, "Navigation prohibited.—Mined areas (1976) may exist along the E section of the Madiq Jubal [the junction between the Gulf of Suez and the Red Sea, 27°46'N, 33°48'E], and an area N of Râs Abu Bakr, bound on the W by the shore . . . Mariners should use caution when transiting near these areas..." (Hydrographic Center, 1978, p. 21).

Offshore areas that contain explosive ordnance which are not shown on hydrographic charts or described in piloting instructions are, of course, even more insidious than the recorded areas. The danger often becomes a problem during dredging, trawling, mineral or oil exploitation, and salvage operations. Some random examples follow. Not too long ago, a World War II sea mine was located on the bottom and had to be removed from the approaches to the port of Le Havre, France (Robertson, 1974). Various World War II bombs have been found over the years in the waterways of West Berlin and neutralized (Associated Press, 1977). Along with a sunken World War II German submarine, nine torpedoes were located some 37 kilometres off the coast of North Carolina, USA, which recently had to be rendered safe (Walker, 1981). A 'Liberty' ship, which sank during World War II with some 3 000 tonnes of explosives on her, continues to lie in the Thames estuary, United Kingdom, off Sheerness (Hogben, 1983). This wreck has been the subject of numerous diver surveys and many learned reports, but its safe disposition remains to be decided upon.

In July 1916 a major explosion occurred in the Port of New York at the Black Tom Ammunition Loading Depot on the New Jersey shore near the Statue of Liberty (Scientific American, 1916). A number of ships and barges blew up. There was heavy loss of life among the stevedores. Shells and other items of ammunition were scattered all about the depot, both on land and underwater. Explosive ordnance sanitation was impossible and the area could never be certified as safe for normal operations. Over the years this part of the Port of New York came to be used as a 'graveyard' or dump for derelict scows and other obsolete craft. It was a terrible eyesore which defied clean-up and rehabilitation. Only in the past decade has the entire area been bulkheaded off and filled in with rock and soil so as to permanently bury the remaining dangerous ammunition. On top of the area is now a park and recreation ground.

Recently, deepwater pipeline laying across the Norwegian trench in the North Sea had to be abruptly suspended because of the unexpected discovery of a number of World War I sea mines on the ocean floor (Vielvoye, 1984). Ten of these uncharted mines had to be destroyed before the operation could be resumed.

III.   On-site analysis

Locating, identifying and determining the amount of hazardous items in an area cannot be done with absolute accuracy. The reasons for this include that: (*a*) duds are not reported or else are reported inaccurately; (*b*) premature and sympathetic detonations are seldom recorded accurately, if at all; (*c*) aircraft and submarine weapons, mines set adrift, floating mines, and artillery projectiles often come to rest far from their intended target; (*d*) official records are often inaccurate or unavailable; and (*e*) the seas, since time immemorial, have been used as the dumping ground for unwanted hazardous material. Recognizing that the available information will be incomplete, the clearance planner should have an established methodology by which reasonable analysis of the problem can be made. The following should be taken into account:

1. That the offshore area may contain sensitive, unexploded shells and other ordnance, both on and buried below the ocean floor, and that it may also contain sea mines on the ocean floor and in the water column.
2. That the tidal area may contain anti-tank, anti-personnel, and amphibious mines.
3. That the casings of the explosive remnants may be more or less badly deteriorated.
4. That the explosive filler may be highly sensitive.
5. That official records and eyewitness reports may be unreliable.
6. That the season, time and weather may interfere with the operation.

The personnel conducting any search and clearance operation must be qualified through training and experience. The local political situation must be favourable and there must be co-operation from the appropriate military and civil authorities as well as from the local population.

It is important to restate that each event involving the explosive remnants of war has both obvious and subtle differences that must be carefully evaluated by trained and experienced personnel. For example, some geographical areas have a so-called weather window during which work can be accomplished, while at other times it is impossible to do so. Inhibiting weather factors include high winds, heavy seas, dust or sand storms, and rain, snow or ice.

The clearance planner must structure the operation so as to 'prove the negative', that is, prove by the thoroughness of the proposed search that all hazardous items would have been detected; and, by the thoroughness of the disposal operation, that all explosive items have been disposed of. Whereas an area, once contaminated, can never be certified absolutely free of hazardous remnants, a high degree of confidence in the clearing can nonetheless be achieved, depending upon the technology applied and the techniques employed.

A typical on-site analysis team might be formed of individuals having the following qualifications: (*a*) expertise in underwater explosive ordnance and its disposal; (*b*) expertise in explosive effects (usually a structural engineer); (*c*) expertise in mine countermeasures; (*d*) expertise in salvage (usually a naval architect); and (*e*) expertise in operations analysis and statistics.

Finally, it is not possible to emphasize strongly enough the necessity for an on-site analysis, one that is undertaken early on by a fully qualified clearance planning team. Indeed, the safety, cost and ultimate success of any clearance operation are directly dependent upon the quality of the planning.

IV. Elementary training

The military forces in most countries undergo various levels of training in explosive ordnance disposal (so-called EOD). Such training ranges from the simple recognition of known ordnance items to advanced courses in the disposal of bombs, mines, and so forth. An individual can never be over-trained for the explosive item to be rendered safe.

In the case of recognition training (sometimes referred to as 'explosive ordnance reconnaissance' or EOR) the individual (either military or civilian) receives instruction in: (*a*) the general characteristics and identification of explosive ordnance; (*b*) the evacuation of people from a hazardous area; (*c*) the marking and control of the area; and (*d*) the method of assisting explosive ordnance disposal personnel.

Training aids are of great importance. They include posters, transparencies (diapositive slides), films and inert or dummy ordnance items. Inert training items are especially important, for use both in the classroom and in the practical field exercises.

Most items of explosive ordnance studied in the classroom are sure to appear more or less different under field conditions, making field training a necessity. Great caution must be taken in carrying out such exercises in order to avoid accidents, which—unfortunately—are known to be all too common.

It will be useful to suggest what is involved in the elementary training of personnel. Such training should begin with an initial two-week course on explosive ordnance, the objective of which is to train selected personnel to locate such hazards, identify them, take appropriate immediate precautions and report to the proper authorities. An 80-hour curriculum for such an initial course might be divided as follows: (*a*) introduction (1 hour); (*b*) basics of explosives (3 hours); (*c*) identification of explosive ordnance (20 hours); (*d*) safety precautions (12 hours); (*e*) on-scene evaluation and organization (3 hours); (*f*) identification exercises in the field (24 hours); (*g*) basic demolition procedures (12 hours); (*h*) final examination (3 hours); and (*i*) conclusion (graduation, etc.) (1 hour).

The necessary follow-up course, of at least two weeks' duration, would be in mine location procedures. Here the curriculum must include the use of mine detectors as well as diver training and other underwater procedures.

Candidates for these courses should be carefully selected since their eventual performance may cause their own or another person's serious injury or death. As a minimum, the trainee should be of above-average intelligence, in good physical condition and comfortable in a remote outdoor environment. The candidate must have no history of mental disorder, must be even-tempered, and should be capable of speaking, reading and writing the language in which the training will be conducted. And he or she should be a cautious, but not timid, person.

It must be noted that the training outlined above is minimal for so-called search-and-alert operations. Moreover, it would not qualify those who completed such training to cope alone with explosive remnants.

V.   Initial clearance

Various sea-mine countermeasures (often referred to as MCM systems) have been developed by the military for the initial clearance of a sea lane or other ocean area. The associated vehicles or platforms include fixed-wing and rotary-wing aircraft, hovercraft and surface ships. Three countermeasure systems might be singled out as examples. The 'Troika' system of the Federal Republic of Germany uses a command-and-control ship to direct three unmanned minesweepers to clear sea lanes of acoustic and magnetic influence mines (Pretty, 1984–1985, p. 205). The French 'Pap' system uses a remotely controlled television-equipped vehicle to either place an explosive charge next to a sea mine which had been previously located by mine-hunting sonar or else to cut its mooring cable (Pretty, 1984–1985, p. 203). The Italian 'Min' system also makes use of a remotely controlled television-equipped vehicle (Pretty, 1984–1985, pp. 209–210). One will appreciate that sea-mine countermeasure operations are best undertaken remotely for reasons of personnel safety.

Numerous passes must be undertaken during minesweeping operations, particularly against influence-minefields, in order to account for or overcome protective circuits in the mines. Mines that rise to the surface as a consequence of mechanical minesweeping are often dispatched by gunfire. Regardless of the countermeasure employed, test ships specially outfitted to withstand underwater explosions (referred to as 'sheep') are normally employed next in order to evaluate whether safe passage has, in fact, been achieved. However, even though the test ship conducts several uneventful transits—and normal surface traffic is permitted to resume—this procedure still does not ensure that all mines have been located and removed from the ocean bottom.

VI. Complete clearance

Following the completion of an initial sea-mine countermeasure operation in a mined area, or after a period of years, it can be assumed that the danger from moored mines and from battery-operated mines has largely ended. The former have been swept, broken from their moorings, sunk to the ocean floor or detonated; and the batteries of the latter are no longer functioning. On the other hand, mines dependent upon contact or another mechanical fuse may still be functional, dud explosive ordnance may continue to be a hazard on the ocean floor, and the high-explosive charges of mines whose fuses no longer function nevertheless remain sensitive and capable of detonation.

Planning

With the determination that a surface ship can safely operate in an area, one must still consider whether or not to attempt to totally clear or sanitize the area in question. Such planning involves the determination of which systems and instruments to employ for bottom searches. It must be noted that the electronic instruments involved are undergoing rapid development.

Sonar is the most common search tool. High-frequency sonar is very good for detecting items protruding from the ocean floor. Under ideal conditions, such sonar sometimes also provides clues to items hidden in the mud. However, low-frequency sonar is better for detecting items beneath the surface of the ocean floor, despite its poorer resolution.

Many different sonar configurations are available. These differ in frequency, pulse length, repetition rate of the pulse and power output. For example, those with a high frequency, short pulse and rapid repetition rate are well suited for the detection of small objects. Sonars can be mounted on the hull of a ship or be towed, and they can be at a fixed or variable depth. The device used most widely in the offshore oil industry for bottom surveys is the towed variable-depth side-scan sonar, of which a number of commercial models are available. Such a side-scan sonar prints a chart record of the ocean floor on each side of the towed apparatus (called a 'fish'). The target contact will appear as a dark mark with a light shadow behind it, from which its size can be estimated. A skilled operator can identify some of the objects recorded, and signal-enhancing computer link-ups are available to assist in such identification. It is also possible to produce a map of the ocean floor that depicts all of the items revealed by the sonar scans. One must recognize, however, that some objects will be transparent to the sound energy emitted by the sonar being used. It is thus good practice to carry out preliminary runs with various test targets to test for transparency.

Magnetic detection instruments, often employed by the offshore oil industry to locate well-heads, pipelines and lost tools below the mud line, are also used to search for underwater ordnance. A variety of magnetic detectors are available, ranging from the older 'flux-gate' types to the newest 'proton-precession' types. Magnetic detectors can be towed singly or in multiple arrays by ships or helicopters. Chart-recording equipment should be used.

Visual search systems—involving humans directly (as divers or in submersibles) or with the help of closed-circuit televisions—have been used for various purposes in the offshore oil industry. Such visual systems are also useful, and sometimes vital, in searching for underwater ordnance. Of course, visual search systems require an adequate degree of underwater visibility. On the other hand, the diver has the additional capability of feel. Divers are more versatile than closed-circuit television systems, but circumstances may well make the employment of divers too dangerous or otherwise impractical.

Remotely controlled vehicles have made great advances in recent years and are taking over many tasks previously performed by divers. Most of these vehicles are equipped with closed-circuit television and sonar and have some form of propulsion. However, total reliance should not be placed on them. Definitive reliance is best placed on a final visual inspection by divers.

A navigation system with an accuracy of plus or minus three metres is an absolute requirement for any underwater search. Excellent commercial systems are available which are portable and which operate from a variety of power supplies. These are necessary in order to make usable the results of the search-sensor recordings described earlier, permitting a return to the target for final identification and disposal.

The support ship—known as the minehunter—should be considered with great care. It must be able to serve as a combination search platform and explosive-ordnance disposal system. It must be capable of serving both as a stable towing vessel and as a diving platform. It should have a low magnetic signature. Offshore workboats, including minehunters, are usually 50–60 metres in length, but are occasionally in the 60–70 metre class for operation in higher seas.

Search, clearance and disposal

Following the planning stage, the operational scenario of the marine explosive-ordnance disposal operation can be divided into the following three phases: (*a*) search, in which 'search' sensors are used to locate and record contacts; (*b*) classification, in which 'classification' sensors are used to investigate and identify all contacts; and (*c*) disposal, in which explosive ordnance is usually eliminated by detonation, and other dangerous debris is removed.

In the search phase, each contact should be recorded, including all pertinent physical information, and a tentative classification assigned. In the classification phase, sonar may also be of use, but visual systems are more reliable, divers more so than systems depending on closed-circuit television. Following refinement of the classification, priorities are established for the disposal phase.

Operations analysis should be employed throughout the clearance operation, that is, in the planning, search, classification and disposal phases. The operations analyst will require a desktop computer with a peripheral track plotter and printer. Especially valuable are calculations of search-effectiveness probability. A calculation of this sort takes into account the capabilities of the sensors being employed, the features of the local environment, the accuracy of the navigation system, the tracking capabilities of the search platform, the quality of the classification procedures, and so forth. The application of operations analysis has been shown to greatly improve both the speed and reliability of a clearance operation, and thus reduce its overall cost as well.

The underwater features in the area to be cleared, both natural and anthropogenic, must be determined during the planning phase and confirmed early in the search phase. The extent of underwater piers, pipelines, moorings, shipwrecks, and so forth can greatly complicate a clearance operation.

The disposal phase has in the past been almost exclusively within the military province. However, as industrial utilization of the ocean expands and encompasses former war zones, the disposal of explosive and other hazardous remnants is often being carried out within the civil sector. Such is becoming the case in the exploration of offshore oil and gas leases, where wells are to be drilled and pipelines laid, and also where hydraulic placer mining is to be practised or ports are to be expanded. It is useful to note here that civil operations are often less complicated and safer than military ones. This is so because they need not be carried out with the speed that a military situation may demand; and because there is no need to retrieve explosive ordnance in an intact condition for intelligence purposes, as can be the case in a military context.

The selection of a disposal technique depends on many variables. Where the surrounding area can withstand a detonation, explosive ordnance is disposed of by countercharging. The countercharges can be placed by divers or, in some cases, by remotely controlled vehicles. The situation may permit charges being lowered from the surface into the proximity of the explosive remnants. In other cases a larger charge may be employed to blow up a whole underwater structure or shipwreck. The size of the total detonation (i.e., countercharge plus the items being disposed of) must be carefully considered owing not only to human safety factors, but also to the potential danger of pollution, ecological damage, or structural damage to nearby installations. In some rare instances, where the explosive remnants are too close to permanent installations or other important features, removal or in-place disarming by divers may be mandatory. Structural engineers who specialize in the technology of shock effects are vital in making such determinations.

VII. Conclusion

During wartime, and especially during the heat of battle, the military requirements for rendering safe a land or ocean area are far more restricted than are the post-war civil requirements. The neutralization of mines and other explosive ordnance in wartime usually involves only the limited clearance of specific lanes, be they land or sea, in order to serve the tactical requirements of the moment. On the other hand, when the war is over and the time comes for the return of these hazardous areas to general productive application and other normal civil pursuits, the need for total sanitization usually becomes necessary.

Post-war clearance in the aquatic environment can be an especially difficult and dangerous undertaking. Such operations often involve the disposal of sea mines; of sunken (and possibly even booby-trapped) barges, landing craft or ships containing cargoes of diverse live ordnance; of jettisoned ammunition; of dud shells, torpedoes and depth charges on the ocean floor; and so forth. The operations must be carried out both with the safety of the disposal personnel and the protection of the local environment in mind. It is hoped that the present chapter has suggested the dimensions of this fearsome legacy of war.

References

Aldridge, J. L. 1980. *Red Sea and Gulf of Aden pilot*. 12th ed. Taunton, UK: Hydrographer of the Navy Publication No. N.P. 64, 284 pp. + pl. + charts.

Associated Press. 1977. Bomb hunt ends in the waterways of West Berlin. *International Herald Tribune*, Paris, **1977** (26–27 Mar): 5.

Hogben, R. 1983. It's that wreck again. *Fairplay International Shipping Weekly*, London, **287**(5215): inside back cover.

Hydrographic Center. 1978. *Sailing directions (en route) for the Red Sea and the Persian Gulf*. Washington: US Department of Defense, Defense Mapping Agency Hydrographic Center Publication No. 172.

Pretty, R. T. (ed.). 1984–1985. *Jane's weapon systems*. 15th ed. London: Jane's Publishing Co., 1017 pp.

Robertson, N. 1974. Immediate end of service by the *France* is ordered. *New York Times* **1974** (19 Sep): 85.

Scientific American. 1916. Munitions explosion in New York harbor: a great disaster accompanied with comparatively little loss of life. *Scientific American*, New York, **115**(7): 150, 161.

Searle, W. F., Jr & Moody, D. H. 1981. *Clearance of explosive ordnance, sea mines and other war debris from the marine environment*. Geneva: UN Inst. for Training & Research Publ. No. UNITAR/EUR/81/WR/18, 35 pp.

Vielvoye, R. 1984. Hazards of pipelaying. *Oil & Gas Journal*, Tulsa, **82**(18): 27.

Walker, P. B. 1981. Disposal of ordnance aboard the sunken German submarine, U-352: finding of no significant environmental impact. *Federal Register*, Washington, **46**: 18757–18758.

7. Explosive remnants of war: detection through the use of dogs

Robert E. Lubow

University of Tel Aviv

I. Introduction

This chapter describes the value of using dogs for detecting the explosive remnants of war. An explanation is given in general terms of how dogs detect explosive munitions in the field, and both their strengths and their limitations are noted. Recommendations are offered which would allow for the rapid and effective utilization of dogs to detect explosive remnants in the immediate post-war environment. The importance is pointed out of creating and maintaining an information bank of relevant research programmes as well as of organizations that maintain and employ explosive-detecting dogs.

Some examples of the successful application of dogs for the detection of explosive munitions are drawn from the experience of the present author and his colleagues during the Second Indochina War (Carr-Harris & Thal, 1970; Lubow, 1977, pp. 173–202). A general history of the employment of animals for military purposes is presented elsewhere (Lubow, 1977).

II. The dog as a detector of explosive munitions

Having developed an overall strategy for searching out explosive munitions, it is necessary to choose the actual detection system that will most effectively accomplish the mission. This choice should be made on the basis of objective criteria with reference to such factors as: (*a*) the time it takes to initiate the search operation; (*b*) the time it takes to complete the operation; (*c*) the reliability to be attained, taking into account both the proportion of munitions overlooked and the number of false alarms; (*d*) the safety of the clearance personnel; and (*e*) the cost of the operation. It is suggested that—if all of the factors are weighed in the balance—the most efficient means for detecting the explosive remnants of war will often be the explosive-detecting dog.

Exact procedures for training dogs (and their handlers) to detect explosive remnants of war have been described elsewhere and are not provided here.[1] It is important to note that such training takes of the order of six months. Moreover, if no facility for this purpose exists, it would take an additional half year or so to first organize the training centre and its staff. Also it should be noted that the team of dog plus handler must be kept in training in order to maintain its skill.

The explosive-detecting dog is a non-aggressive dog; German shepherds (Alsatians) and Labrador retrievers are often used. It is trained to detect and respond to various types of explosive ordnance as well as to the individual components. The primary cues are the vapours that escape from the explosive filler, which the dog detects through its keen and discriminating sense of smell. By a combination of olfactory and visual cues the dog is also trained to respond to the metal or plastic casing and other components as well as to the trip wires used to set off some mines and booby traps.

The dogs are routinely trained to detect munitions buried as deeply as 15 centimetres below the surface, suspended as high as 150 centimetres above the ground, and located as far to the side as 3 metres. Actually, the distance a dog can be from the object in order to make a detection varies considerably with terrain and wind conditions. Under ideal circumstances—that is, with a steady breeze and the dog approaching from the downwind direction—detections can be made from distances of up to 60 metres or so. Under the worst circumstances—that is, with the dog approaching from the upwind direction or under conditions of a brisk cross-wind—the dog may have to get within 30 centimetres.

The dogs are trained so that when they detect the presence of an explosive item they will (unless directed otherwise by the handler) approach to a distance of about 60 centimetres from it and then sit. They are trained to work both on-leash and off-leash, but to maximize their efficiency under most circumstances they should be off the leash. They can then work up to 90 metres or more in front of their handler, frequently out of visual contact. If the dog is equipped with a harness-mounted radio transmitter and the handler with a receiver, then the handler is able to recognize a sit response without seeing the dog. This is useful when a trail or other strip is being cleared that winds through forest, goes over rough terrain, or is in an urban area. In clearing a trail or road, the dog will typically move out at a trot, traversing the pathway from side to side. Its overall speed of forward motion will, of course, vary with the local conditions, but might average 1.5 kilometres per hour. A dog can work in this fashion for about 1–2 hours before requiring a rest, and no more than a total of 5–6 hours in any one day.

Trained dogs are extremely effective in their ability to detect explosive

[1] A substantial investment into the training of explosive-detecting dogs has been made by the USA in recent years and a considerable technical literature on the subject has been made available (e.g., Breland & Bailey, 1971; Carr-Harris & Thal, 1970; Dean, 1972; Mitchell, 1976a; 1976b; 1976c; Nolan & Gravitte, 1977; Romba, 1970; 1974).

munitions via their olfactory sense. Indeed, their ability to do so exceeds that of existing electronic detectors, including those based on sensing the vapours given off by the explosive. Unfortunately, as emplaced munitions age they emit decreasing amounts of odour. Thus, the ability of dogs to detect explosive ordnance decreases with time (Breland & Bailey, 1971). For example, the detection of anti-personnel mines (US model M-14) which had been experimentally emplaced 16 months previously was found to be seriously impaired. This shortcoming—a serious one—is shared by electronic detectors of non-metallic mines, but not by the metal detectors used for metallic mines and duds.

Lessons from the Second Indochina War

US armed forces carried out two operational field trials with explosive-detecting dogs during the Second Indochina War. In the first of these the US Army brought a dog platoon to South Viet Nam in 1969 for a period of six months which included about 14 dogs and their handlers previously trained for this purpose (White, 1969).

These explosive-detecting dogs were employed in a number of ways. For example, on a patrol mission, when the unit moved out, a dog was made to assume the point position. The handler would follow the dog about 15 metres to the rear, flanked by two riflemen. This relatively secure position permitted the handler to focus the necessary attention on the behaviour of the dog. The use of dogs in this fashion was deemed a success because mines and booby traps discovered by the dog could then be safely bypassed by the patrol.

An operation designed to clear a road of mines was performed somewhat more elaborately. A dog and its handler, or perhaps two dogs plus their handlers, were again made the lead element. A dog would often work some 45 metres in front of its handler. The dog/handler team or teams were frequently followed by two teams each equipped with a metal detector. These in turn were occasionally followed by a vehicle pushing heavy, sand-filled steel rollers ahead of it, meant to detonate any buried pressure-sensitive mines which had been missed. As much as nine kilometres of road could be cleared in this fashion during a strenuous seven-hour work day (Romba, 1970).

During the course of this trial operation, the dogs detected a total of 76 explosive munitions (either the object itself or its trip wire), but overlooked 12 others subsequently detected by other means (Romba, 1970). Several of the misses were of ordnance that had been in place for a long period of time and two occurred after heavy rains. Two of the misses were also missed by the metal-detector teams. Several of the dogs were wounded in action, but none was killed.

In the second of the operational field trials, the US Marine Corps brought 14 explosive-detecting dogs plus handlers to South Viet Nam in 1970 for a period of two months. During one 12-day period these dog/handler teams detected eight emplaced (buried) mines variously fitted with pressure fuses or trip wires, plus four buried dud shells and grenades. The evaluators concluded that the dogs could locate mines, whether standard or improvised, and that the dogs were operationally suitable for supplementing other detection measures.

III. Conclusion

It is taken for granted that an international body should be established, presumably under the auspices of the United Nations, that can be called upon by a country following a war in order to obtain advice and assistance in the clearance of the explosive remnants of that war. It is additionally taken for granted that explosive-detecting dogs would be an extraordinarily useful component of any such post-war clearance operation.

In order to minimize the deployment time of dogs, agreements should be reached between the envisioned international agency and a number of institutions that maintain facilities for training explosive-detecting dogs so that their services can be called upon as needed. In the absence of, or in addition to, such agreements the international agency itself should perhaps establish and maintain its own dog-training facility.[2] At the very least, an information bank should be created and maintained that contains the details of all active training facilities for, and users of, explosive-detecting dogs. Then, even in the absence of an international dog facility or of pre-arranged agreements there would be the opportunity for the international agency to quickly negotiate an agreement for the rapid deployment of dogs. Such a plan would mitigate one of the major drawbacks in the use of dogs in detecting explosive remnants, their loss in efficiency with time.

References

Breland, M. & Bailey, R. E. 1971. *Specialized mine detector dog.* Aberdeen Proving Ground, Maryland: US Army Land Warfare Laboratory Technical Memorandum No. LWL-CR-04B70, 16 pp. (AD-736 860).

Carr-Harris, E. & Thal, R. 1970. *Mine, booby-trap, tripwire and tunnel detection: final report.* Raleigh, N. Carolina: Behavior Systems Report No. LWL-CR02B67 (LWL-CR01B68), 74 pp. (AD-867 404/6).

Customs Co-operation Council. 1976. Dogs detecting drugs. *Bulletin on Narcotics*, New York, **28**(3): 41–60.

Dean, E. E. 1972. *Feasibility study on training infantry multipurpose dogs: final report.* San Antonio, Texas: Southwest Research Institute Report No. LWL-CR-06B70, 35 pp. (AD-746 998).

[2] If under United nations auspices, an enlarged dog-training centre might well deal not only with the detection of explosive remnants of war, but also with the detection of battlefield cadavers as well as with the detection of trapped casualties (alive or dead) of such natural disasters as earthquakes. A further role for such a centre might be for the detection of narcotics which are in the process of being smuggled (Customs Co-operation Council, 1976).

Lubow, R. E., 1977. *War animals*. Garden City, NY: Doubleday, 255 pp.

Mitchell, D. S. 1976a. *Selection of dogs for land mine and booby trap detection: final technical report; volume I*. San Antonio, Texas: Southwest Research Institute, 61 pp. (AD-A031 980/6GA).

Mitchell, D. S. 1976b. *Training and employment of land mine and booby trap detector dogs: final technical report: volume II*. San Antonio, Texas: Southwest Research Institute, 247 pp. (AD-A031 981/4GA).

Mitchell, D. S. 1976c. *User's guide: land mine and booby trap detector dogs: final technical report: volume III*. San Antonio, Texas: Southwest Research Institute, 67 pp. (AD-A031 982/2GA).

Nolan, R. V. & Gravitte, D. L. 1977. *Mine-detecting canines*. Fort Belvoir, Virginia: US Army Mobility Equipment Research and Development Command Report No. MERADCOM-2217, 83 pp. (AD-A048 748/8GA).

Romba, J. J. 1970. *Tactics in the development of mine detector dogs*. Aberdeen Proving Ground, Maryland: US Army Land Warfare Laboratory, 8 pp. (AD-713 577).

Romba, J. J. 1974. *Study in training methodology of mine dogs: final report*. Aberdeen Proving Ground, Maryland: US Army Land Warfare Laboratory Report No. LWL-TR-74-91, 14 pp. (AD-784 048/1GA).

White, B. O., Jr. 1969. *60th Infantry Platoon (Scout dogs) (Mine/tunnel detector dogs): final report*. Saigon: US Army Concept Team in Vietnam, 42 pp. (AD-869 383/0).

8. Explosive remnants of war: legal aspects

Jozef Goldblat
Stockholm International Peace Research Institute

I. Introduction

The material remnants of war include a variety of high-explosive munitions, which have not detonated either because they malfunctioned at the time of their delivery to the target (so-called duds), or because they were not activated following emplacement. This paper deals in particular with the second category of munitions—that is, mines and booby traps—which had been laid in order to injure or deter the enemy, but which were neither triggered nor removed, thus remaining potentially lethal.

Both mines and booby traps are explicitly referred to in the existing rules of international law. These rules are analysed here, and the need for reinforcing their implementation is examined from the point of view of protecting the human environment. Related reviews and analyses of the laws of war are available elsewhere (Goldblat, 1982b; 1983; see also appendix 1).

II. Prohibitions and restrictions

Since mines and booby traps can be employed both for offensive and defensive purposes, their absolute prohibition outside the context of general disarmament has never been seriously considered. However, their use is regulated by the general principles of international customary law, which are supplemented by specific norms of the conventional humanitarian law of armed conflict.

The pertinent basic principles of customary law are those prohibiting the use of weapons which indiscriminately affect both combatants and non-combatants, as well as the resort to methods of warfare that cause superfluous injury or unnecessary suffering. Especially relevant are the principles regarding the protection of the civilian population, including those prohibiting action expected to cause incidental losses or injuries, as well as damage to civilian objects, or denial of objects which are indispensable for the survival of the population.

In addition, rules of conventional law specifically restrict and limit the types and missions of mines and booby traps, as well as the places where they might lawfully be used.

Types of munition subject to restriction

The use of both sea and land mines is regulated by international multilateral treaties. Thus, the Hague Convention VIII of 1907 (see appendix 5) forbids the laying of unanchored submarine mines (to avoid carrying the danger beyond the scene of conflict), except when they are so constructed as to become harmless one hour at most after the person who laid them ceases to control them. The one-hour exception was made to preserve the right of a war vessel which was being pursued to drop off free mines in order to delay or destroy its pursuers (Levie, 1971–1972). Also forbidden is the use of submarine mines which do not become harmless as soon as they have broken loose from their moorings, as well as torpedoes which do not become harmless when they have missed their mark. (These would be dangerous not only to the enemy, but also to third countries, and even to the user himself.)

The provisions of the 1907 Hague Convention VIII refer only to "automatic contact mines", one of the two types in existence at the turn of the century, the other being mines for close-in protection of harbours, electrically controlled and detonated from a shore facility, and comparatively non-controversial. Nevertheless, considering the motives of self-interest of the belligerents which lay behind the adoption of this Convention, there is no good reason why the subsequently developed acoustic and magnetic mines, which can be set off by noise or by changes in the magnetic field, should not fall under the same restrictions [apropos this, see appendix 6].

Protocol II of the Inhumane Weapons Convention of 1981 (see appendix 3) relates to the use on land of mines, booby traps and "other devices" (defined collectively as manually emplaced munitions actuated by remote control or automatically after a lapse of time). Protocol II makes a distinction between "mine" and "remotely delivered mine": the first means any munition placed under, on, or near the ground or other surface area and designed to be detonated or exploded by the presence, proximity, or contact of a person or vehicle; and the second means any mine delivered by artillery, rocket, mortar, or similar means or dropped from an aircraft. Booby traps are devices which function when a person disturbs or approaches an apparently harmless object or performs an apparently safe act. All these weapons are subject to restrictions in populated areas, whereas the use of remotely delivered mines is prohibited, unless a self-actuating or remotely controlled mechanism is used to render each mine harmless, or to destroy it, when it no longer serves the military purpose for which it was placed in position.

The Seabed Treaty of 1971 (Goldblat, 1982a, pp. 175–177) prohibits emplanting or emplacing on the seabed and the ocean floor, and in the subsoil thereof, beyond the outer limit of a 22 kilometre (12 nautical mile) seabed zone any nuclear weapons or any other type of weapon of mass destruction. Implicit in this prohibition is the proscription of nuclear mines, as well as mines containing chemical and biological warfare agents, anchored to or installed on the sea bottom.

Prohibited missions

Under the Hague Convention VIII of 1907, automatic contact mines may not be laid with the sole object of intercepting commercial shipping. In introducing a subjective element, this provision has created some ambiguity, as it is not possible to prove that the mines have no military objective. Under Protocol II of the Inhumane Weapons Convention of 1981, the placement of the weapons specified therein, which is not on or directed against a military objective, or the employment of a method or means of delivery which cannot be directed at a specific military objective, is prohibited. The prohibitions and restrictions of this Protocol do not apply to the use of anti-ship mines at sea or in inland waterways, but they do apply to mines laid to interdict beaches, waterway crossings or river crossings.

Geographic extent of prohibitions and restrictions

Under the Hague Convention VIII of 1907, it is forbidden to lay automatic contact mines off the coast and ports of the enemy with the sole object of intercepting commercial shipping. There is no express prohibition on the use of mines on the high seas. However, one can argue that because it would deny, at least temporarily, the rights of navigation to neutrals, mining of the high seas would be in conflict with the principle of the freedom of the high seas.

Protocol II of the Inhumane Weapons Convention of 1981 prohibits the use of mines, booby traps, and other devices in any city, town, village, or other area containing a similar concentration of civilians, in which combat between ground forces is not taking place or does not appear to be imminent, unless they are placed on or close to a military objective under the control of an adverse party.

III. Precautionary measures during hostilities

Under the Hague Convention VIII of 1907, the belligerents shall take every possible precaution for the security of peaceful shipping. They are to do "their utmost" to render their mines harmless within a limited time and, should these cease to be under surveillance, notification is to be given of the danger zones to ship owners and governments as soon as military exigencies permit. Neutral powers laying mines off their coasts must observe the same rules.

Protocol II of the Inhumane Weapons Convention of 1981 also requires that all "feasible" precautions (defined as "practicable or practically possible", taking into account all circumstances ruling at the time) should be taken to protect civilians from the effects of mines and booby traps. The precautionary measures could include, for example, the posting of warning signs, the posting of sentries, the issuance of warnings or the provision of fences. Advance warning shall be given of any delivery or dropping of remotely delivered mines which may affect the civilian population "unless circumstances do not permit".

IV.   Precautionary measures after hostilities

Under the Hague Convention VIII of 1907, the parties are obliged to do "their utmost" to remove the mines which they have laid, each removing its own mines. With regard to mines laid by one of the belligerents off the coast of the other, their position must be made known to the other party by the power which laid them, and each power must proceed "with the least possible delay" to remove the mines in its own waters.

According to Protocol II of the Inhumane Weapons Convention of 1981, "immediately" after the cessation of active hostilities, all information concerning the location of minefields, mines, and booby traps must be made available in order to protect civilians. (The locations are to be recorded by the parties during the hostilities in conformity with the agreed guidelines.) Special protection from the effects of mines and booby traps is to be provided to a United Nations force or mission performing the functions of peace-keeping or observation. And finally, the parties should "endeavour" to reach agreement, both among themselves and with other states and international organizations, on providing information and technical and material assistance—including, in appropriate circumstances, joint oper-ations—necessary to remove or otherwise render ineffective minefields, mines and booby traps placed in position during the conflict.

A recent example of a negotiated commitment to mine clearance is provided by the Viet Nam–US Protocols of 1973 concerning the removal of the explosive remnants of war (see appendix 7) accompanying the Paris Agreement ending the Second Indochina War, under which the USA undertook *inter alia* to remove, deactivate or destroy all the mines in the territorial waters, ports, harbours and waterways of North Viet Nam.

V.   Breaches and responsibility

The main sanction for violation of the law applicable to armed conflict is international opprobrium and condemnation. In this respect, the rules related to mines and booby traps are no exception.

There remains, however, the question of material responsibility for damages caused by the remnants of war, as well as for their removal. Although the laying of mines and booby traps is not in itself an act which entails international responsibility, article 3 of the Hague Convention IV of 1907 (Goldblat, 1982a, pp. 122–124) established a principle, subsequently reiterated in article 91 of the 1977 Protocol I (Goldblat, 1982a, pp. 239–252) additional to the 1949 Geneva Conventions for the protection of war victims, that a party to the conflict violating the provisions of the 1907 Convention or the 1977 Protocol shall be liable to pay compensation. The mentioned Protocol provides for an international fact-finding commission to enquire into facts alleged to be a "grave" breach.

However, the establishment of breaches in the case of mine or booby-trap laying could present insurmountable difficulties, among other reasons, because the provisos attached to the prohibitions and restrictions admit a subjective qualification of military necessity. It would be unrealistic to expect that some judicial authority could make rulings concerning the interpretation and application of international humanitarian law that would be binding on states (Sandoz, 1981). In any event, the question of reimbursement can hardly be solved piecemeal for each category of weapon or method of war wrongfully used. It is related to the considerably broader problem of war reparations, which have been conceived of as a sanction against an aggressor, but which have so far been exacted only by victors from the vanquished under post-war settlements.

Damages could also be brought about unintentionally, both during a war and as a result of it, even if all the rules had been observed. The opinion that compensation for the harmful consequences of the remnants of war can be claimed not only on the ground of the illegality of the act committed, but also on the basis of the very fact of the damage caused (Blishchenko, 1981), that is, also in the absence of wrongful intent ("objective" or "strict" liability) is not widely shared. It would not be right to draw direct analogies between the damages caused in time of peace, which are clearly subject to compensation, as in the case of damages caused by space objects (under the 1972 Convention of International Liability), and those damages that are a result of war, which may or may not be. One cannot overlook that wars are waged precisely for the purpose of injuring the enemy. There can be no question of responsibility for damages done if the generally accepted rules of conduct in armed confict are observed. (Defining responsibility for aggression is a separate problem.) But since the continued presence of mines and booby traps after the cessation of hostilities serves no useful military purpose, it would seem more practical, and also more in accord with the humanitarian aspects of the law of armed conflict, to remedy concrete situations where human lives are threatened and where economic recovery and development are obstructed, than to try to apportion blame for the sake of retribution.

In many cases, rehabilitation of the environment cannot be achieved by the belligerents themselves. It will be noted that the Geneva Convention III of 1949 (see appendix 4) prohibits the use of prisoners of war, against their will, in the removal of mines or similar devices. Financial resources and the technical competence for effective mine and booby-trap clearance may require the help of third countries. The postulate of international co-operation in dealing with international problems of a humanitarian character can be found in article 1 of the United Nations Charter. Moreover, the 1970 United Nations 'Declaration of Principles of International Law concerning Friendly Relations and Co-operation among States in accordance with the Charter of the United Nations' (UNGA, 1970) which expresses the consensus of the international community as to the legal principles inherent in the United Nations Charter, makes it clear that member states of the United Nations have a "duty" to co-operate with one another, irrespective of the differences in their political, economic and social systems, in the various spheres of international relations. This implies that precedence should be given to humanitarian imperatives over military and political considerations, in particular, over the determination of guilt for starting the war or for non-observance of the laws. In other words, the interests of the civilian populations must be safeguarded both in the aggressor and the victim states.

VI. Conclusion

As mentioned above, the Inhumane Weapons Convention of 1981 provides for *ad hoc* arrangements, to be agreed upon after the cessation of hostilities, in order to remove or render ineffective the material remnants of war. However, given the general interest in the speedy setting in motion and completion of the relevant operations, it would appear advisable to establish a standing mechanism in time of peace, which could be called upon at short notice to render services upon the termination of hostilities, or even during the war in areas lying outside the actual battle zone.

References

Blishchenko, I. P. 1981. *Legal basis of claims for damage to the environment resulting from military action*. Geneva: UN Inst. for Training & Research Publ. No. UNITAR/EUR/81/WR/1, 30 pp.

Goldblat, J. 1982a. *Agreements for arms control: a critical survey*. London: Taylor & Francis, 388 pp. [a SIPRI book].

Goldblat, J. 1982b. Laws of armed conflict: an overview of the restrictions and limitations on the methods and means of warfare. *Bulletin of Peace Proposals*, Oslo, **13**: 127–133.

Goldblat, J. 1983. Convention on 'inhumane' weapons. *Bulletin of the Atomic Scientists*, Chicago, **39**(1): 24–25.

Levie, H. S. 1971–1972. Mine warfare and international law. *Naval War College Review*, Newport, Rhode Island, **24**(8): 27–35.

Sandoz, Y. 1981. *Unlawful damage in armed conflicts and reparation therefor under international humanitarian law.* Geneva: UN Inst. for Training & Research Publ. No. UNITAR/EUR/81/WR/7, 35 pp.

UNGA (United Nations General Assembly). 1970. *Declaration of principles of international law concerning friendly relations and co-operation among states in accordance with the charter of the United Nations.* New York: United Nations General Assembly Resolution No. 2625 (XXV) (24 Oct 1970), 12 pp. Reprinted in: *UN Yearbook*, New York, **24**: 788–792.

Appendix 1.   Explosive remnants of war: select bibliography

Arthur H. Westing
Stockholm International Peace Research Institute

Fenrick, W. J. 1981. New developments in the law concerning the use of conventional weapons in armed conflict. *Canadian Yearbook of International Law*, Vancouver, **19**: 229–256.

Foss, C. F. (ed.). 1984–1985. *Jane's armour and artillery.* 5th ed. London: Jane's Publishing Co., 897 pp.

Foss, C. F. & Gander, T. J. (eds). 1984. *Jane's military vehicles and ground support equipment.* 5th ed. London: Jane's Publishing Co., 871 pp.

Goad, K. J. W. & Halsey, D. H. J. 1982. *Ammunition (including grenades and mines).* Oxford: Brassey's Publishers, 289 pp.

Hartmann, G. K. 1979. *Weapons that wait: mine warfare in the U.S. Navy.* Annapolis, Maryland: Naval Institute Press, 294 pp.

Hogg, I. V. (ed.). 1984–1985. *Jane's infantry weapons.* 10th ed. London: Jane's Publishing Co., 957 pp.

Levie, H. S. 1971–1972. Mine warfare and international law. *Naval War College Review*, Newport, Rhode Island, **24**(8): 27–35.

Libya. 1981. *White book: some examples of the damages caused by the belligerents of the World War II to the people of the Jamahiriya.* Tripoli: Libyan Studies Centre, 176 pp.

Lubow, R. E. 1977. *War animals.* Garden City, NY: Doubleday, 255 pp.

Lumsden, M. 1978. *Anti-personnel weapons.* London: Taylor & Francis, 299 pp. [a SIPRI book].

Mostofi, A. 1983. *Remnants of war.* New York: UN Inst. for Training & Research Publ. No. UNITAR/CR/26, 124 pp.

Owen, J. (ed.). 1979. *Brassey's infantry weapons of the world.* 2nd ed. London: Brassey's Publishers, 480 pp.

Partsch, K. J. 1984. Remnants of war as a legal problem in the light of the Libyan case. *American Journal of International Law*, Washington, **78**: 386–401.

Pretty, R. T. (ed.). 1984–1985. *Jane's weapon systems*. 15th ed. London: Jane's Publishing Co., 1017 pp.

Red Cross, International Committee of the. 1973. *Weapons that may cause unnecessary suffering or have indiscriminate effects*. Geneva: Intl Committee of the Red Cross, 72 pp.

Tresckow, A.v. 1975. [Land mines.] (In German). *Soldat und Technik*, Frankfurt a.M., **18**: 388–400.

Truver, S. C. 1985. Mines of August: an international whodunit. *United States Naval Institute Proceedings*, Annapolis, Maryland, **111**(5): 94–117.

Westing, A. H. 1984. Remnants of war. *Ambio*, Stockholm, **13**: 14–17.

Appendix 2.   Explosive remnants of war: a chronology of United Nations activities, 1975–1984[1]

1975

1.   UNGA by Resolution 3435(XXX) of 9 December 1975 *inter alia* recognized that the development of certain developing countries was impeded by material remnants of war, especially mines, called upon the states who took part in those wars to make available all relevant information, and requested UNEP to undertake a study of the problem.

1976

1.   UNEP's Executive Director presented to UNEP recommendations for such a study (UNEP/GC/84/Add.1; 24 March 1976).

2.   UNEP by Decision 80(IV) of 9 April 1976 *inter alia* authorized its Executive Director to carry out this study, to consult with governments regarding the feasibility of an inter-governmental meeting on the subject, and to render relevant assistance to those states that request it.

3.   UNEP's Executive Director presented to UNGA an interim study report (A/31/210; 13 September 1976).

4.   UNGA by Resolution 31/111 of 16 December 1976 *inter alia* requested UNEP to complete the study.

1977

1.   UNEP's Executive Director presented to UNEP a study report (UNEP/GC/103; 19 April 1977).

2.   UNEP by Decision 101(V) of 25 May 1977 *inter alia* requested its Executive Director to continue to pursue the matters of an inter-governmental meeting and the rendering of assistance to states upon request.

[1] The United Nations is herein referred to as 'UN'; the United Nations General Assembly (New York) as 'UNGA'; the United Nations Environment Programme (Nairobi) as 'UNEP'; the United Nations Institute for Training and Research (New York) as 'UNITAR'; and the Stockholm International Peace Research Institute as 'SIPRI'.

3. UNEP presented to UNGA its study report (A/32/137; 27 July 1977).

4. UNGA by Resolution 32/168 of 19 December 1977 *inter alia* took note of UNEP's report and invited concerned governments to co-operate in the matter with UNEP's Executive Director.

1978

1. UNEP's Executive Director reported to UNEP on the general lack of interest or reticence of governments in convening an inter-governmental meeting on the subject (UNEP/GC.6/18; 2 February 1978 plus Add. 1; 5 May 1978).

2. UNEP by Decision 6/15 of 15 May 1978 *inter alia* requested its Executive Director to urge all governments to provide relevant information in their possession, to render assistance upon request, and to continue to study the problem.

1979

Apparently no relevant UN action during this year.

1980

1. UNGA by Resolution 35/71 of 5 December 1980 *inter alia* regretted that no real action had as yet been taken on the subject, again called upon states to make available relevant information, and suggested the possibility of convening a conference under UN auspices.

1981

1. UNGA presided over the signing of the Inhumane Weapons Convention of 10 April 1981, Protocol II of which regulates the employment of land mines and booby traps (see appendix 3).

2. UNITAR convened a symposium on the material remnants of World War II in Libya, held in Geneva 28 April–1 May 1981 (UNITAR/EUR/81/WR/1–23; 1981 and UNITAR/CR/26; 1983).

3. UNEP by Decision 9/5 of 25 May 1981 again appealed to states to supply available relevant information and requested its Executive Director to co-operate with UNGA's Secretary-General regarding an international conference on the subject.

4. UNEP's secretariat presented to UNGA a progress report (A/36/531; 25 September 1981).

5. UNGA by Resolution 36/188 of 17 December 1981 *inter alia* took note of UNEP's progress report and reiterated its request for a conference.

1982

1.   UNEP by Decision 10/8 of 28 May 1982 *inter alia* again supported the idea of an international conference.

2.   UNGA's Secretary-General presented to UNGA a report on the subject (A/37/415; 23 September 1982).

3.   UNGA by Resolution 37/215 of 20 December 1982 *inter alia* requested its Secretary-General to co-operate with UNEP's Executive Director in preparing a factual study on the subject, including its economic, environmental and legal aspects; and again to consider the possibility of convening a conference.

1983

1.   UNEP entered into a contract with SIPRI, effective 1 February 1983, for the latter *inter alia* to review the environmental aspects of the subject and to organize a related technical UNEP expert group meeting.

2.   UNEP (with the assistance of SIPRI) convened an expert group meeting on explosive remnants of conventional war, held in Geneva 25–28 July 1983 which resulted in a report to UNEP (A/38/383/annex; 19 October 1983) (see appendix 8).

3.   UNGA's Secretary-General presented to UNGA a report on the subject (A/38/383; 19 October 1983).

4.   UNGA by Resolution 38/162 of 19 December 1983 *inter alia* took note of its Secretary's-General report, again regretted a lack of concrete action by governments, endorsed the recommendations of the report to UNEP (see appendix 8, section VIII), and requested the Secretary-General, in co-operation with the Executive Director of UNEP, to submit a progress report to its next session.

1984

1.   UNGA's Secretary-General presented to UNGA a brief report on the subject (A/39/580; 15 October 1984).

2.   UNGA by Resolution 39/167 of 17 December 1984 *inter alia* took note of its Secretary's-General report, again regretted a lack of concrete action, and again requested the Secretary-General to submit a report to its next session.

Appendix 3.   Inhumane Weapons Convention of 1981

I. Text

The Convention on Prohibitions or Restrictions on the Use of certain
Conventional Weapons which may be Deemed to be Excessively Injurious
or to have Indiscriminate Effects was signed at New York on 10 April 1981
and (the United Nations Secretary-General, the Depositary, having received
the requisite 20 ratifications) entered into force on 2 December 1983. The
parties to the Convention are given in section II below. The text of the
Convention follows (Goldblat, 1982, pp. 296–302):

The High Contracting Parties,

Recalling that every State has the duty, in conformity with the Charter of the United
Nations, to refrain in its international relations from the threat or use of force against
the sovereignty, territorial integrity or political independence of any State, or in any
other manner inconsistent with the purposes of the United Nations,

Further recalling the general principle of the protection of the civilian population
against the effects of hostilities,

Basing themselves on the principle of international law that the right of the parties to
an armed conflict to choose methods or means of warfare is not unlimited, and on the
principle that prohibits the employment in armed conflicts of weapons, projectiles and
material and methods of warfare of a nature to cause superfluous injury or
unnecessary suffering,

Also recalling that it is prohibited to employ methods or means of warfare which are
intended, or may be expected, to cause widespread, long-term and severe damage to
the natural environment,

Confirming their determination that in cases not covered by this Convention and its
annexed Protocols or by other international agreements, the civilian population and
the combatants shall at all times remain under the protection and authority of the
principles of international law derived from established custom, from the principles of
humanity and from the dictates of public conscience,

Desiring to contribute to international détente, the ending of the arms race and the
building of confidence among States, and hence to the realization of the aspiration of
all peoples to live in peace,

Recognizing the importance of pursuing every effort which may contribute to
progress towards general and complete disarmament under strict and effective international control,

Reaffirming the need to continue the codification and progressive development of the rules of international law applicable in armed conflict,

Wishing to prohibit or restrict further the use of certain conventional weapons and believing that the positive results achieved in this area may facilitate the main talks on disarmament with a view to putting an end to the production, stockpiling and proliferation of such weapons,

Emphasizing the desirability that all States become parties to this Convention and its annexed Protocols, especially the militarily significant States,

Bearing in mind that the General Assembly of the United Nations and the United Nations Disarmament Commission may decide to examine the question of a possible broadening of the scope of the prohibitions and restrictions contained in this Convention and its annexed Protocols,

Further bearing in mind that the Committee on Disarmament may decide to consider the question of adopting further measures to prohibit or restrict the use of certain conventional weapons,

Have agreed as follows:

Article 1. Scope of application

This Convention and its annexed Protocols shall apply in the situations referred to in Article 2 common to the Geneva Conventions of 12 August 1949 for the Protection of War Victims,[1] including any situation described in paragraph 4 of Article 1 of Additional Protocol I to these Conventions.[2]

Article 2. *Relations with other international agreements*

Nothing in this Convention or its annexed Protocols shall be interpreted as detracting from other obligations imposed upon the High Contracting Parties by international humanitarian law applicable in armed conflict.

Article 3. Signature

This Convention shall be open for signature by all States at United Nations Headquarters in New York for a period of twelve months from 10 April 1981.

Article 4. Ratification, acceptance, approval or accession

1. This Convention is subject to ratification, acceptance or approval by Signatories.

[1] Geneva Conventions of 1949, article 2: In addition to the provisions which shall be implemented in peacetime, the present Convention shall apply to all cases of declared war or of any other armed conflict which may arise between two or more of the High Contracting Parties, even if the state of war is not recognized by one of them.

The Convention shall also apply to all cases of partial or total occupation of the territory of a High Contracting Party, even if the said occupation meets with no armed resistance.

Although one of the Powers in conflict may not be a party to the present Convention, the Powers who are parties thereto shall remain bound by it in their mutual relations. They shall furthermore be bound by the Convention in relation to the said Power, if the latter accepts and applies the provisions thereof.

[2] Geneva Conventions of 1949, Additional Protocol I of 1977, article 1.4: The situations referred to in the preceding paragraph include armed conflicts in which peoples are fighting against colonial domination and alien occupation and against racist régimes in the exercise of their right of self-determination, as enshrined in the Charter of the United Nations and the Declaration on Principles of International Law concerning Friendly Relations and Co-operation among States in accordance with the Charter of the United Nations.

Any State which has not signed this Convention may accede to it.

2. The instruments of ratification, acceptance, approval or accession shall be deposited with the Depositary.

3. Expressions of consent to be bound by any of the Protocols annexed to this Convention shall be optional for each State, provided that at the time of the deposit of its instrument of ratification, acceptance or approval of this Convention or of accession thereto, that State shall notify the Depositary of its consent to be bound by any two or more of these Protocols.

4. At any time after the deposit of its instrument of ratification, acceptance or approval of this Convention or of accession thereto, a State may notify the Depositary of its consent to be bound by any annexed Protocol by which it is not already bound.

5. Any Protocol by which a High Contracting Party is bound shall for that Party form an integral part of this Convention.

Article 5. Entry into force

1. This Convention shall enter into force six months after the date of deposit of the twentieth instrument of ratification, acceptance, approval or accession.

2. For any State which deposits its instrument of ratification, acceptance, approval or accession after the date of the deposit of the twentieth instrument of ratification, acceptance, approval or accession, this Convention shall enter into force six months after the date on which that State has deposited its instrument of ratification, acceptance, approval or accession.

3. Each of the Protocols annexed to this Convention shall enter into force six months after the date by which twenty States have notified their consent to be bound by it in accordance with paragraph 3 or 4 of Article 4 of this Convention.

4. For any State which notifies its consent to be bound by a Protocol annexed to this Convention after the date by which twenty States have notified their consent to be bound by it, the Protocol shall enter into force six months after the date on which that State has notified its consent so to be bound.

Article 6. Dissemination

The High Contracting Parties undertake, in time of peace as in time of armed conflict, to disseminate this Convention and those of its annexed Protocols by which they are bound as widely as possible in their respective countries and, in particular, to include the study thereof in their programmes of military instruction, so that those instruments may become known to their armed forces.

Article 7. Treaty relations upon entry into force of this Convention

1. When one of the parties to a conflict is not bound by an annexed Protocol, the parties bound by this Convention and that annexed Protocol shall remain bound by them in their mutual relations.

2. Any High Contracting Party shall be bound by this Convention and any Protocol annexed thereto which is in force for it, in any situation contemplated by Article 1, in relation to any State which is not a party to this Convention or bound by the relevant annexed Protocol, if the latter accepts and applies this Convention or the relevant Protocol, and so notifies the Depositary.

3. The Depositary shall immediately inform the High Contracting Parties concerned of any notification received under paragraph 2 of this Article.

4. This Convention, and the annexed Protocols by which a High Contracting Party is

bound, shall apply with respect to an armed conflict against that High Contracting Party of the type referred to in Article 1, paragraph 4, of Additional Protocol I to the Geneva Conventions of 12 August 1949 for the Protection of War Victims:

    (*a*) where the High Contracting Party is also a party to Additional Protocol I and an authority referred to in Article 96, paragraph 3, of that Protocol has undertaken to apply the Geneva Conventions and Additional Protocol I in accordance with Article 96, paragraph 3, of the said Protocol, and undertakes to apply this Convention and the relevant annexed Protocols in relation to that conflict; or

    (*b*) where the High Contracting Party is not a party to Additional Protocol I and an authority of the type referred to in subparagraph (*a*) above accepts and applies the obligations of the Geneva Conventions and of this Convention and the relevant annexed Protocols in relation to that conflict. Such an acceptance and application shall have in relation to that conflict the following effects:

        (i) the Geneva Conventions and this Convention and its relevant annexed Protocols are brought into force for the parties to the conflict with immediate effect;

        (ii) the said authority assumes the same rights and obligations as those which have been assumed by a High Contracting Party to the Geneva Conventions, this Convention and its relevant annexed Protocols; and

        (iii) the Geneva Conventions, this Convention and its relevant annexed Protocols are equally binding upon all parties to the conflict.

The High Contracting Party and the authority may also agree to accept and apply the obligations of Additional Protocol I to the Geneva Conventions on a reciprocal basis.

Article 8. Review and amendments

    1. (*a*) At any time after the entry into force of this Convention any High Contracting Party may propose amendments to this Convention or any annexed Protocol by which it is bound. Any proposal for an amendment shall be communicated to the Depositary, who shall notify it to all the High Contracting Parties, and shall seek their views on whether a conference should be convened to consider the proposal. If a majority, that shall not be less than eighteen of the High Contracting Parties so agree, he shall promptly convene a conference to which all the High Contracting Parties shall be invited. States not parties to this Convention shall be invited to the conference as observers.

        (*b*) Such a conference may agree upon amendments which shall be adopted and shall enter into force in the same manner as this Convention and the annexed Protocols, provided that amendments to this Convention may be adopted only by the High Contracting Parties and that amendments to a specific annexed Protocol may be adopted only by the High Contracting Parties which are bound by that Protocol.

    2. (*a*) At any time after the entry into force of this Convention any High Contracting Party may propose additional protocols relating to other categories of conventional weapons not covered by the existing annexed protocols. Any such proposal for an additional protocol shall be communicated to the Depositary, who shall notify it to all the High Contracting Parties in accordance with subparagraph 1(*a*) of this Article. If a majority, that shall not be less than eighteen of the High Contracting Parties so agree, the

Depositary shall promptly convene a conference to which all States shall be invited.

(b) Such a conference may agree, with the full participation of all States represented at the conference, upon additional protocols which shall be adopted in the same manner as this Convention, shall be annexed thereto and shall enter into force as provided in paragraphs 3 and 4 of Article 5 of this Convention.

3. (a) If, after a period of ten years following the entry into force of this Convention, no conference has been convened in accordance with subparagraph 1(a) or 2(a) of this Article, any High Contracting Party may request the Depositary to convene a conference to which all High Contracting Parties shall be invited to review the scope and operation of this Convention and the Protocols annexed thereto and to consider any proposal for amendments of this Convention or of the existing Protocols. States not parties to this Convention shall be invited as observers to the conference. The conference may agree upon amendments which shall be adopted and enter into force in accordance with subparagraph 1(b) above.

(b) At such conference consideration may also be given to any proposal for additional protocols relating to other categories of conventional weapons not covered by the existing annexed Protocols. All States represented at the conference may participate fully in such consideration. Any additional protocols shall be adopted in the same manner as this Convention, shall be anexed thereto and shall enter into force as provided in paragraphs 3 and 4 of Article 5 of this Convention.

(c) Such a conference may consider whether provision should be made for the convening of a further conference at the request of any High Contracting Party if, after a similar period to that referred to in subparagraph 3(a) of this Article, no conference has been convened in accordance with subparagraph 1(a) or 2(a) of this Article.

Article 9. Denunciation

1. Any High Contracting Party may denounce this Convention or any of its annexed Protocols by so notifying the Depositary.

2. Any such denunciation shall only take effect one year after receipt by the Depositary of the notification of denunciation. If, however, on the expiry of that year the denouncing High Contracting Party is engaged in one of the situations referred to in Article 1, the Party shall continue to be bound by the obligations of this Convention and of the relevant annexed Protocols until the end of the armed conflict or occupation and, in any case, until the termination of operations connected with the final release, repatriation or re-establishment of the persons protected by the rules of international law applicable in armed conflict, and in the case of any annexed Protocol containing provisions concerning situations in which peace-keeping, observation or similar functions are performed by United Nations forces or missions in the area concerned, until the termination of those functions.

3. Any denunciation of this Convention shall be considered as also applying to all annexed Protocols by which the denouncing High Contracting Party is bound.

4. Any denunciation shall have effect only in respect of the denouncing High Contracting Party.

5. Any denunciation shall not affect the obligations already incurred, by reason of an armed conflict, under this Convention and its annexed Protocols by such denouncing High Contracting Party in respect of any act committed before this denunciation becomes effective.

Article 10. Depositary

1. The Secretary-General of the United Nations shall be the Depositary of this Convention and of its annexed Protocols.

2. In addition to his usual functions, the Depositary shall inform all States of:

(*a*) signatures affixed to this Convention under Article 3;

(*b*) deposits of instruments of ratification, acceptance or approval of or accession to this Convention deposited under Article 4;

(*c*) notifications of consent to be bound by annexed Protocols under Article 4;

(*d*) the dates of entry into force of this Convention and of each of its annexed Protocols under Article 5; and

(*e*) notifications of denunciation received under Article 9 and their effective date.

Article 11. Authentic texts

The original of this Convention with the annexed Protocols, of which the Arabic, Chinese, English, Russian and Spanish texts are equally authentic, shall be deposited with the Depositary, who shall transmit certified true copies thereof to all States.

PROTOCOL (I) ON NON-DETECTABLE FRAGMENTS

It is prohibited to use any weapon the primary effect of which is to injure by fragments which in the human body escape detection by X-rays.

PROTOCOL (II) ON PROHIBITIONS OR RESTRICTIONS ON THE USE OF MINES, BOOBY-TRAPS AND OTHER DEVICES

Article 1. Material scope of application

This Protocol relates to the use on land of the mines, booby-traps and other devices defined herein, including mines laid to interdict beaches, waterway crossings or river crossings, but does not apply to the use of anti-ship mines at sea or in inland waterways.

Article 2. Definitions

For the purpose of this Protocol:

1. "Mine" means any munition placed under, on or near the ground or other surface area and designed to be detonated or exploded by the presence, proximity or contact of a person or vehicle, and "remotely delivered mine" means any mine so defined delivered by artillery, rocket, mortar or similar means or dropped from an aircraft.

2. "Booby-trap" means any device or material which is designed, constructed or adapted to kill or injure and which functions unexpectedly when a person disturbs or approaches an apparently harmless object or performs an apparently safe act.

3. "Other devices" means manually-emplaced munitions and devices designed to kill, injure or damage and which are actuated by remote control or automatically after a lapse of time.

4. "Military objective" means, so far as objects are concerned, any object which by its nature, location, purpose or use makes an effective contribution to military action and whose total or partial destruction, capture or neutralization, in the circumstances ruling at the time, offers a definite military advantage.

5. "Civilian objects" are all objects which are not military objectives as defined in paragraph 4.

6. "Recording" means a physical, administrative and technical operation designed to obtain, for the purpose of registration in the official records, all available information facilitating the location of minefields, mines and booby-traps.

Article 3. General restrictions on the use of mines, booby-traps and other devices

1. This Article applies to:

(*a*) mines;

(*b*) booby-traps; and

(*c*) other devices.

2. It is prohibited in all circumstances to direct weapons to which this Article applies, either in offence, defence or by way of reprisals, against the civilian population as such or against individual civilians.

3. The indiscriminate use of weapons to which this Article applies is prohibited. Indiscriminate use is any placement of such weapons:

(*a*) which is not on, or directed against, a military objective; or

(*b*) which employs a method or means of delivery which cannot be directed at a specific military objective; or

(*c*) which may be expected to cause incidental loss of civilian life, injury to civilians, damage to civilian objects, or a combination thereof, which would be excessive in relation to the concrete and direct military advantage anticipated.

4. All feasible precautions shall be taken to protect civilians from the effects of weapons to which this Article applies. Feasible precautions are those precautions which are practicable or practically possible taking into account all circumstances ruling at the time, including humanitarian and military considerations.

Article 4. Restrictions on the use of mines other than remotely delivered mines, booby-traps and other devices in populated areas

1. This Article applies to:

(*a*) mines other than remotely delivered mines;

(*b*) booby-traps; and

(*c*) other devices.

2. It is prohibited to use weapons to which this Article applies in any city, town, village or other area containing a similar concentration of civilians in which combat between ground forces is not taking place or does not appear to be imminent, unless either:

(*a*) they are placed on or in the close vicinity of a military objective belonging to or under the control of an adverse party; or

(*b*) measures are taken to protect civilians from their effects, for example, the posting of warning signs, the posting of sentries, the issue of warnings or the provision of fences.

Article 5. Restrictions on the use of remotely delivered mines

1. The use of remotely delivered mines is prohibited unless such mines are only used within an area which is itself a military objective or which contains military objectives, and unless:

(a) their location can be accurately recorded in accordance with Article 7(1)(a); or

(b) an effective neutralizing mechanism is used on each such mine, that is to say, a self-actuating mechanism which is designed to render a mine harmless or cause it to destroy itself when it is anticipated that the mine will no longer serve the military purpose for which it was placed in position, or a remotely-controlled mechanism which is designed to render harmless or destroy a mine when the mine no longer serves the military purpose for which it was placed in position.

2. Effective advance warning shall be given of any delivery or dropping of remotely delivered mines which may affect the civilian population, unless circumstances do not permit.

Article 6. Prohibition on the use of certain booby-traps

1. Without prejudice to the rules of international law applicable in armed conflict relating to treachery and perfidy, it is prohibited in all circumstances to use:

(a) any booby-trap in the form of an apparently harmless portable object which is specifically designed and constructed to contain explosive material and to detonate when it is disturbed or approached; or

(b) booby-traps which are in any way attached to or associated with:

(i) internationally recognized protective emblems, signs or signals;

(ii) sick, wounded or dead persons;

(iii) burial or cremation sites or graves;

(iv) medical facilities, medical equipment, medical supplies or medical transportation;

(v) children's toys or other portable objects or products specially designed for the feeding, health, hygiene, clothing or education of children;

(vi) food or drink;

(vii) kitchen utensils or appliances except in military establishments, military locations or military supply depots;

(viii) objects clearly of a religious nature;

(ix) historic monuments, works of art or places of worship which constitute the cultural or spiritual heritage of peoples;

(x) animals or their carcasses.

2. It is prohibited in all circumstances to use any booby-trap which is designed to cause superfluous injury or unnecessary suffering.

Article 7. Recording and publication of the location of minefields, mines and booby-traps

1. The parties to a conflict shall record the location of:

(a) all pre-planned minefields laid by them; and

(b) all areas in which they have made large-scale and pre-planned use of booby-traps.

2. The parties shall endeavour to ensure the recording of the location of all other minefields, mines and booby-traps which they have laid or placed in position.

3. All such records shall be retained by the parties who shall:

   (*a*) immediately after the cessation of active hostilities:

      (i) take all necessary and appropriate measures, including the use of such records, to protect civilians from the effects of minefields, mines and booby-traps; and either

     (ii) in cases where the forces of neither party are in the territory of the adverse party, make available to each other and to the Secretary-General of the United Nations all information in their possession concerning the location of minefields, mines and booby-traps in the territory of the adverse party; or

    (iii) once complete withdrawal of the forces of the parties from the territory of the adverse party has taken place, make available to the adverse party and to the Secretary-General of the United Nations all information in their possession concerning the location of minefields, mines and booby-traps in the territory of the adverse party;

   (*b*) when a United Nations force or mission performs functions in any area, make available to the authority mentioned in Article 8 such information as is required by that Article;

   (*c*) whenever possible, by mutual agreement, provide for the release of information concerning the location of minefields, mines and booby-traps, particularly in agreements governing the cessation of hostilities.

Article 8. Protection of United Nations forces and missions from the effects of minefields, mines and booby-traps

   1. When a United Nations force or mission performs functions of peace-keeping, observation or similar functions in any area, each party to the conflict shall, if requested by the head of the United Nations force or mission in that area, as far as it is able:

   (*a*) remove or render harmless all mines or booby-traps in that area;

   (*b*) take such measures as may be necessary to protect the force or mission from the effects of minefields, mines and booby-traps while carrying out its duties; and

   (*c*) make available to the head of the United Nations force or mission in that area, all information in the party's possession concerning the location of minefields, mines and booby-traps in that area.

   2. When a United Nations fact-finding mission performs functions in any area, any party to the conflict concerned shall provide protection to that mission except where, because of the size of such mission, it cannot adequately provide such protection. In that case it shall make available to the head of the mission the information in its possession concerning the location of minefields, mines and booby-traps in that area.

Article 9. International co-operation in the removal of minefields, mines and booby-traps

   After the cessation of active hostilities, the parties shall endeavour to reach agreement, both among themselves and, where appropriate, with other States and with international organizations, on the provision of information and technical and material assistance—including, in appropriate circumstances, joint operations—necessary to remove or otherwise render ineffective minefields, mines and booby-traps placed in position during the conflict.

Technical annex to protocol (II)

Guidelines on recording

Whenever an obligation for the recording of the location of minefields, mines and booby-traps arises under the Protocol, the following guidelines shall be taken into account.

1. With regard to pre-planned minefields and large-scale and pre-planned use of booby-traps:
   (a) maps, diagrams or other records should be made in such a way as to indicate the extent of the minefield or booby-trapped area; and
   (b) the location of the minefield or booby-trapped area should be specified by relation to the co-ordinates of a single reference point and by the estimated dimensions of the area containing mines and booby-traps in relation to that single reference point.

2. With regard to other minefields, mines and booby-traps laid or placed in position: In so far as possible, the relevant information specified in paragraph 1 above should be recorded so as to enable the areas containing minefields, mines and booby-traps to be identified.

PROTOCOL (III) ON PROHIBITIONS OR RESTRICTIONS ON THE USE OF INCENDIARY WEAPONS

Article 1. Definitions

For the purpose of this Protocol:

1. "Incendiary weapon" means any weapon or munition which is primarily designed to set fire to objects or to cause burn injury to persons through the action of flame, heat, or a combination thereof, produced by a chemical reaction of a substance delivered on the target.
   (a) Incendiary weapons can take the form of, for example, flame throwers, fougasses, shells, rockets, grenades, mines, bombs and other containers of incendiary substances.
   (b) Incendiary weapons do not include:
      (i) Munitions which may have incendiary effects, such as illuminants, tracers, smoke or signalling systems;
      (ii) Munitions designed to combine penetration, blast or fragmentation effects with an additional incendiary effect, such as armour-piercing projectiles, fragmentation shells, explosive bombs and similar combined-effects munitions in which the incendiary effect is not specifically designed to cause burn injury to persons, but to be used against military objectives, such as armoured vehicles, aircraft and installations or facilities.

2. "Concentration of civilians" means any concentration of civilians, be it permanent or temporary, such as in inhabited parts of cities, or inhabited towns or villages, or as in camps or columns of refugees or evacuees, or groups of nomads.

3. "Military objective" means, so far as objects are concerned, any object which by its nature, location, purpose or use makes an effective contribution to military action and whose total or partial destruction, capture or neutralization, in the circumstances ruling at the time, offers a definite military advantage.

4. "Civilian objects" are all objects which are not military objectives as defined in paragraph 3.

5. "Feasible precautions" are those precautions which are practicable or practically possible taking into account all circumstances ruling at the time, including humanitarian and military considerations.

Article 2. Protection of civilians and civilian objects

1. It is prohibited in all circumstances to make the civilian population as such, individual civilians or civilian objects the object of attack by incendiary weapons.

2. It is prohibited in all circumstances to make any military objective located within a concentration of civilians the object of attack by air-delivered incendiary weapons.

3. It is further prohibited to make any military objective located within a concentration of civilians the object of attack by means of incendiary weapons other than air-delivered incendiary weapons, except when such military objective is clearly separated from the concentration of civilians and all feasible precautions are taken with a view to limiting the incendiary effects to the military objective and to avoiding, and in any event to minimizing, incidental loss of civilian life, injury to civilians and damage to civilian objects.

4. It is prohibited to make forests or other kinds of plant cover the object of attack by incendiary weapons except when such natural elements are used to cover, conceal or camouflage combatants or other military objectives, or are themselves military objectives.

II. Parties

As of January 1985, the Inhumane Weapons Convention of 1981 has accumulated a total of 22 parties (this sum not including Byelorussia and the Ukraine, both constituent republics of the USSR). Of the five permanent members of the United Nations Security Council, so far only China and the USSR have become parties, whereas France, the United Kingdom and the USA have not. A list of the 22 parties to the Convention as of January 1985, together with their year of joining, follows (Goldblat & Ferm, 1984, pp. 653–676); augmented by information from the Depositary):

Australia (1983), Austria (1983), Bulgaria (1982), China (1982), Czechoslovakia (1982), Denmark (1982), Ecuador (1982), Finland (1982), German DR (1982), Guatemala (1983), Hungary (1982), India (1984), Japan (1982), Laos (1983), Mexico (1982), Mongolia (1982), Norway (1983), Poland (1983), Sweden (1982), Switzerland (1982), USSR (1982), Yugoslavia (1983).

References

Goldblat, J. 1982. *Agreements for arms control: a critical survey*. London: Taylor & Francis, 387 pp. [a SIPRI book].
Goldblat, J. & Ferm, R. 1984. Arms control agreements. *World Armaments and Disarmament, SIPRI Yearbook 1984*. London: Taylor & Francis, pp. 637–676.

Appendix 4.   Geneva Convention III of 1949

I.   Text

The Geneva Convention III Relative to the Treatment of Prisoners of War was signed at Geneva on 12 August 1949 and (Switzerland, the Depositary, having received the requisite two ratifications) entered into force on 21 October 1950. The parties to the Convention are given in section II below. Excerpts from the text of the Convention follow (Roberts & Guelff, 1982, pp. 215–270, 326–337):

The undersigned Plenipotentiaries of the Governments represented at the Diplomatic Conference held at Geneva from April 21 to August 12, 1949, for the purpose of revising the Convention concluded at Geneva on July 27, 1929, relative to the Treatment of Prisoners of War, have agreed as follows:

Article 1

The High Contracting Parties undertake to respect and to ensure respect for the present Convention in all circumstances.

Article 2

In addition to the provisions which shall be implemented in peace time, the present Convention shall apply to all cases of declared war or of any other armed conflict which may arise between two or more of the High Contracting Parties, even if the state of war is not recognized by one of them.

The Convention shall also apply to all cases of partial or total occupation of the territory of a High Contracting Party, even if the said occupation meets with no armed resistance.

Although one of the Powers in conflict may not be a party to the present Convention, the Powers who are parties thereto shall remain bound by it in their mutual relations. They shall furthermore be bound by the Convention in relation to the said Power, if the latter accepts and applies the provisions thereof.

Article 3

In the case of armed conflict not of an international character occurring in the territory of one of the High Contracting Parties, each Party to the conflict shall be bound to apply, as a minimum, the following provisions:

1. Persons taking no active part in the hostilities, including members of armed forces who have laid down their arms and those placed *hors de combat* by sickness, wounds, detention, or any other cause, shall in all circumstances be treated humanely, without any adverse distinction founded on race, colour, religion or faith, sex, birth or wealth, or any other similar criteria.

To this end, the following acts are and shall remain prohibited at any time and in any place whatsoever with respect to the above-mentioned persons:

(*a*) violence to life and person, in particular murder of all kinds, mutilation, cruel treatment and torture;

(*b*) taking of hostages;

(*c*) outrages upon personal dignity, in particular, humiliating and degrading treatment;

(*d*) the passing of sentences and the carrying out of executions without previous judgment pronounced by a regularly constituted court affording all the judicial guarantees which are recognized as indispensable by civilized peoples.

2. The wounded and sick shall be collected and cared for.

An impartial humanitarian body, such as the International Committee of the Red Cross, may offer its services to the Parties to the conflict.

The Parties to the conflict should further endeavour to bring into force, by means of special agreements, all or part of the other provisions of the present Convention.

The application of the preceding provisions shall not affect the legal status of the Parties to the conflict.

Article 4

A. Prisoners of war, in the sense of the present Convention, are persons belonging to one of the following categories, who have fallen into the power of the enemy:

1. Members of the armed forces of a Party to the conflict as well as members of militias or volunteer corps forming part of such armed forces.

2. Members of other militias and members of other volunteer corps, including those of organized resistance movements, belonging to a Party to the conflict and operating in or outside their own territory, even if this territory is occupied, provided that such militias or volunteer corps, including such organized resistance movements, fulfil the following conditions:

(*a*) that of being commanded by a person responsible for his subordinates;

(*b*) that of having a fixed distinctive sign recognizable at a distance;

(*c*) that of carrying arms openly;

(*d*) that of conducting their operations in accordnce with the laws and customs of war.

3. Members of regular armed forces who profess allegiance to a government or an authority not recognized by the Detaining Power.

4. Persons who accompany the armed forces without actually being members thereof, such as civilian members of military aircraft crews, war correspondents, supply contractors, members of labour units or of services responsible for the welfare of the armed forces, provided that they have received authorization from the armed forces which they accompany, who shall provide them for that purpose with an identity card similar to the annexed model.

5. Members of crews, including masters, pilots and apprentices, of the merchant marine and the crews of civil aircraft of the Parties to the conflict, who do not benefit

by more favourable treatment under any other provisions of international law.

6. Inhabitants of a non-occupied territory, who on the approach of the enemy spontaneously take up arms to resist the invading forces, without having had time to form themselves into regular armed units, provided they carry arms openly and respect the laws and customs of war.

B. The following shall likewise be treated as prisoners of war under the present Convention:

1. Persons belonging, or having belonged, to the armed forces of the occupied country, if the occupying Power considers it necessary by reason of such allegiance to intern them, even though it has originally liberated them while hostilities were going on outside the territory it occupies, in particular where such persons have made an unsuccessful attempt to rejoin the armed forces to which they belong and which are engaged in combat, or where they fail to comply with a summons made to them with a view to internment.

2. The persons belonging to one of the categories enumerated in the present Article, who have been received by neutral or non-belligerent Powers on their territory and whom these Powers are required to intern under international law, without prejudice to any more favourable treatment which these Powers may choose to give and with the exception of Articles 8, 10, 15, 30, fifth paragraph, 58–67, 92, 126 and, where diplomatic relations exist between the Parties to the conflict and the neutral or non-belligerent Power concerned, those Articles concerning the Protecting Power. Where such diplomatic relations exist, the Parties to a conflict on whom these persons depend shall be allowed to perform towards them the functions of a Protecting Power as provided in the present Convention, without prejudice to the functions which these Parties normally exercise in conformity with diplomatic and consular usage and treaties.

C. This Article shall in no way affect the status of medical personnel and chaplains as provided for in Article 33 of the present Convention.

Article 5

The present Convention shall apply to the persons referred to in Article 4 from the time they fall into the power of the enemy and until their final release and repatriation.

Should any doubt arise as to whether persons, having committed a belligerent act and having fallen into the hands of the enemy, belong to any of the categories enumerated in Article 4, such persons shall enjoy the protection of the present Convention until such time as their status has been determined by a competent tribunal.

Article 6

In addition to the agreements expressly provided for in Articles 10, 23, 28, 33, 60, 65, 66, 67, 72, 73, 75, 109, 110, 118, 119, 122 and 132, the High Contracting Parties may conclude other special agreements for all matters concerning which they may deem it suitable to make separate provision. No special agreement shall adversely affect the situation of prisoners of war, as defined by the present Convention, nor restrict the rights which it confers upon them.

Prisoners of war shall continue to have the benefit of such agreements as long as the Convention is applicable to them, except where express provisions to the contrary are contained in the aforesaid or in subsequent agreements, or where more favourable measures have been taken with regard to them by one or other of the Parties to the conflict.

Article 7

Prisoners of war may in no circumstances renounce in part or in entirety the rights secured to them by the present Convention, and by the special agreements referred to in the foregoing Article, if such there be.

• • •

Article 52

Unless he be a volunteer, no prisoner of war may be employed on labour which is of an unhealthy or dangerous nature.

No prisoner of war shall be assigned to labour which would be looked upon as humiliating for a member of the Detaining Power's own forces.

The removal of mines or similar devices shall be considered as dangerous labour.

• • •

Article 136

The present Convention, which bears the date of this day, is open to signature until February 12, 1950, in the name of the Powers represented at the Conference which opened at Geneva on April 21, 1949; furthermore, by Powers not represented at that Conference, but which are parties to the Convention of July 27, 1929.

Article 137

The present Convention shall be ratified as soon as possible and the ratifications shall be deposited at Berne.

A record shall be drawn up of the deposit of each instrument of ratification and certified copies of this record shall be transmitted by the Swiss Federal Council to all the Powers in whose name the Convention has been signed, or whose accession has been notified.

Article 138

The present Convention shall come into force six months after not less than two instruments of ratification have been deposited.

Thereafter, it shall come into force for each High Contracting Party six months after the deposit of the instrument of ratification.

Article 139

From the date of its coming into force, it shall be open to any Power in whose name the present Convention has not been signed, to accede to this Convention.

Article 140

Accessions shall be notified in writing to the Swiss Federal Council, and shall take effect six months after the date on which they are received.

The Swiss Federal Council shall communicate the accessions to all the Powers in whose name the Convention has been signed, or whose accession has been notified.

Article 141

The situations provided for in Articles 2 and 3 shall give immediate effect to ratifications deposited and accessions notified by the Parties to the conflict before or after the beginning of hostilities or occupation. The Swiss Federal Council shall communicate by the quickest method any ratifications or accessions received from Parties to the conflict.

Article 142

Each of the High Contracting Parties shall be at liberty to denounce the present Convention.

The denunciation shall be notified in writing to the Swiss Federal Council, which shall transmit it to the Governments of all the High Contracting Parties.

The denunciation shall take effect one year after the notification thereof has been made to the Swiss Federal Council. However, a denunciation of which notification has been made at a time when the denouncing Power is involved in a conflict shall not take effect until peace has been concluded, and until after operations connected with the release and repatriation of the persons protected by the present Convention have been terminated.

The denunciation shall have effect only in respect of the denouncing Power. It shall in no way impair the obligations which the Parties to the conflict shall remain bound to fulfil by virtue of the principles of the law of nations, as they result from the usages established among civilized peoples, from the laws of humanity and the dictates of the public conscience.

Article 143

The Swiss Federal Council shall register the present Convention with the Secretariat of the United Nations. The Swiss Federal Council shall also inform the Secretariat of the United Nations of all ratifications, accessions and denunciations received by it with respect to the present Convention.

In witness whereof the undersigned, having deposited their respective full powers, have signed the present Convention.

Done at Geneva this twelfth day of August 1949, in the English and French languages. The original shall be deposited in the Archives of the Swiss Confederation. The Swiss Federal Council shall transmit certified copies thereof to each of the signatory and acceding States.

• • •

II. Parties

As of January 1985, the Geneva Convention III of 1949 has accumulated a total of 158 parties (this sum not including Byelorussia and the Ukraine, both constituent republics of the USSR, and not Namibia, which is not an independent state). Of the five permanent members of the United Nations Security Council, all have become parties, that is, China, France, the United Kingdom, the USA and the USSR. A list of the 158 parties to the Convention as of January 1985, together with the year of joining, follows (Roberts & Guelff, 1982, pp. 326–330; augmented by information from the Depositary):

Afghanistan (1956), Albania (1957), Algeria (1960), Angola (1984), Argentina (1956), Australia (1958), Austria (1953), Bahamas (1975), Bahrain (1971), Bangladesh (1972), Barbados (1968), Belgium (1952), Belize (1984), Benin (1961), Bolivia (1976), Botswana (1968), Brazil (1957), Bulgaria (1954), Burkina Faso (1961), Burundi (1971), Cameroon (1963), Canada (1965), Cape Verde (1984), Central African Rep. (1966), Chad (1970), Chile (1950), China (1956), Colombia (1961), Congo (1967), Costa Rica (1969), Cuba (1954), Cyprus (1962), Czechoslovakia (1950), Denmark (1951), Djibouti (1978), Dominica (1981), Dominican Rep. (1958), Ecuador (1954), Egypt (1952), El Salvador (1953), Ethiopia (1969), Fiji (1971), Finland (1955), France (1951), Gabon (1965), Gambia (1966), German DR (1956), Germany, FR (1954), Ghana (1958), Greece (1956), Grenada (1981), Guatemala (1952), Guinea (1984), Guinea-Bissau (1974), Guyana (1968), Haiti (1957), Honduras (1965), Hungary (1954), Iceland (1965), India (1950), Indonesia (1958), Iran (1957), Iraq (1956), Ireland (1962), Israel (1951), Italy (1951), Ivory Coast (1961), Jamaica (1964), Japan (1953), Jordan (1951), Kampuchea (1958), Kenya (1966), Korea, DPR (1957), Korea, Rep. (1966), Kuwait (1967), Laos (1956), Lebanon (1951), Lesotho (1968), Liberia (1954), Libya (1956), Liechtenstein (1950), Luxembourg (1953), Madagascar (1963), Malawi (1968), Malaysia (1962), Mali (1965), Malta (1968), Mauritania (1962), Mauritius (1970), Mexico (1952), Monaco (1950), Mongolia (1958), Morocco (1956), Mozambique (1983), Nepal (1964), Netherlands (1954), New Zealand (1959), Nicaragua (1953), Niger (1964), Nigeria (1961), Norway (1951), Oman (1974), Pakistan (1951), Panama (1956), Papua New Guinea (1976), Paraguay (1961), Peru (1956), Philippines (1952), Poland (1954), Portugal (1961), Qatar (1975), Romania (1954), Rwanda (1964), St Lucia (1981), St Vincent & Grenadines (1981), Samoa (1984), San Marino (1953), Sao Tomé & Principe (1976), Saudi Arabia (1963), Senegal (1963), Seychelles (1984), Sierra Leone (1965), Singapore (1973), Solomon Islands (1981), Somalia (1962), South Africa (1952), Spain (1952), Sri Lanka (1959), Sudan (1957), Suriname (1976), Swaziland (1973), Sweden (1953), Switzerland (1950), Syria (1953), Tanzania (1962), Thailand (1954), Togo (1962), Tonga (1978), Trinidad & Tobago (1963), Tunisia (1957), Turkey (1954), Tuvalu (1981), Uganda (1964), United Arab Emirates (1972), United Kingdom (1957), Uruguay (1969), USA (1955), USSR (1954), Vanuatu (1982), Vatican City (1951), Venezuela (1956), Viet Nam (1957), Yemen Arab Rep. (1970), Yemen, PDR (1977), Yugoslavia (1950), Zaïre (1961), Zambia (1966), Zimbabwe (1983).

Reference

Roberts, A. & Guelff, R. (eds). 1982. *Documents on the laws of war.* Oxford: Clarendon Press, 498 pp.

Appendix 5.   Hague Convention VIII of 1907

I.   Text

The Hague Convention VIII Relative to the Laying of Automatic Submarine Contact Mines was signed at the Hague on 18 October 1907 and (the Netherlands, the Depositary, having received the requisite two ratifications) entered into force on 26 January 1910. The parties to the Convention are given in section II below. The text of the Convention follows (Friedman, 1972, pp. 342–347, 271–277):

[The representatives of 38 Powers]:

Inspired by the principle of the freedom of sea routes, the common highway of all nations;

Seeing that, although the existing position of affairs makes it impossible to forbid the employment of automatic submarine contact mines, it is nevertheless desirable to restrict and regulate their employment in order to mitigate the severity of war and to ensure, as far as possible, to peaceful navigation the security to which it is entitled, despite the existence of war;

Until such time as it is found possible to formulate rules on the subject which shall ensure to the interests involved all the guarantees desirable;

Have resolved to conclude a Convention for this purpose, and . . . have agreed upon the following provisions:

Article 1

It is forbidden:

1. To lay unanchored automatic contact mines, except when they are so constructed as to become harmless one hour at most after the person who laid them ceases to control them;

2. To lay anchored automatic contact mines which do not become harmless as soon as they have broken loose from their moorings;

3. To use torpedoes which do not become harmless when they have missed their mark.

Article 2

It is forbidden to lay automatic contact mines off the coast and ports of the enemy, with the sole object of intercepting commercial shipping.

Article 3

When anchored automatic contact mines are employed, every possible precaution must be taken for the security of peaceful shipping.

The belligerents undertake to do their utmost to render these mines harmless within a limited time, and, should they cease to be under surveillance, to notify the danger zones as soon as military exigencies permit, by a notice addressed to ship owners, which must also be communicated to the Governments through the diplomatic channel.

Article 4

Neutral Powers which lay automatic contact mines off their coasts must observe the same rules and take the same precautions as are imposed on belligerents.

The neutral Power must inform ship owners, by a notice issued in advance, where automatic contact mines have been laid. This notice must be communicated at once to the Governments through the diplomatic channel.

Article 5

At the close of the war, the contracting Powers undertake to do their utmost to remove the mines which they have laid, each Power removing its own mines.

As regards anchored automatic contact mines laid by one of the belligerents off the coast of the other, their position must be notified to the other party by the Power which laid them, and each Power must proceed with the least possible delay to remove the mines in its own waters.

Article 6

The contracting Powers which do not at present own perfected mines of the pattern contemplated in the present Convention, and which, consequently, could not at present carry out the rules laid down in Articles 1 and 3, undertake to convert the *matériel* of their mines as soon as possible, so as to bring it into conformity with the foregoing requirements.

Article 7

The provisions of the present Convention do not apply except between contracting Powers, and then only if all the belligerents are parties to the Convention.

Article 8

The present Convention shall be ratified as soon as possible.

The ratifications shall be deposited at The Hague.

The first deposit of ratifications shall be recorded in a *procès-verbal* signed by the representatives of the Powers which take part therein and by the Netherland Minister for Foreign Affairs.

The subsequent deposits of ratifications shall be made by means of a written notification addressed to the Netherland Government and accompanied by the instrument of ratification.

A duly certified copy of the *procès-verbal* relative to the first deposit of ratifications, of the notifications mentioned in the preceding paragraph, as well as of the instruments of ratification, shall be at once sent, by the Netherland Government, through the diplomatic channel, to the Powers invited to the Second Peace Conference, as well as to the other Powers which have adhered to the Convention. In the cases contemplated in the preceding paragraph, the said Government shall inform them at the same time of the date on which it has received the notification.

Article 9

Non-signatory Powers may adhere to the present Convention.

The Power which desires to adhere notifies in writing its intention to the Netherland Government, transmitting to it the act of adhesion, which shall be deposited in the archives of the said Government.

This Government shall at once transmit to all the other Powers a duly certified copy of the notification as well as of the act of adhesion, stating the date on which it received the notification.

Article 10

The present Convention shall come into force, in the case of the Powers which were a party to the first deposit of ratifications, sixty days after the date of the *procès-verbal* of this deposit, and, in the case of the Powers which ratify subsequently or adhere, sixty days after the notification of their ratification or of their adhesion has been received by the Netherland Government.

Article 11

The present Convention shall remain in force for seven years, dating from the sixtieth day after the date of the first deposit of ratifications.

Unless denounced, it shall continue in force after the expiration of this period.

The denunciation shall be notified in writing to the Netherland Government, which shall at once communicate a duly certified copy of the notification to all the Powers, informing them of the date on which it was received.

The denunciation shall only have effect in regard to the notifying Power, and six months after the notification has reached the Netherland Government.

Article 12

The contracting Powers undertake to reopen the question of the employment of automatic contact mines six months before the expiration of the period contemplated in the first paragraph of the preceding article, in the event of the question not having been already reopened and settled by the Third Peace Conference.

If the contracting Powers conclude a fresh Convention relative to the employment of mines, the present Convention shall cease to be applicable from the moment it comes into force.

Article 13

A register kept by the Netherland Ministry for Foreign Affairs shall give the date of the deposit of ratifications made in virtue of Article 8, paragraphs 3 and 4, as well as the date on which the notifications of adhesion (Article 9, paragraph 2) or of denunciation (Article 11, paragraph 3) have been received.

Each contracting Power is entitled to have acces to this register and to be supplied with duly certified extracts from it.

In faith whereof the plenipotentiaries have appended their signatures to the present Convention.

Done at The Hague, the 18th October, 1907, in a single copy, which shall remain deposited in the archives of the Netherland Government, and duly certified copies of which shall be sent, through the diplomatic channel, to the Powers which have been invited to the Second Peace Conference.

II. Parties

As of January 1985, the Hague Convention VIII of 1907 has accumulated a total of 27 parties. Of the five permanent members of the United Nations Security Council, so far China, France (with a substantive reservation), the United Kingdom (with a substantive reservation), and the USA have become parties, whereas the USSR has not. A list of all 27 parties to the Convention as of January 1985, together with their year of joining, follows (Roberts & Guelff, 1982, pp. 90–91):

Austria (1909), Belgium (1910), Brazil (1914), China (1917), Denmark (1909), El Salvador (1909), Ethiopia (1935), Fiji (1973), Finland (1918), France (under reservation of article 2; 1910), Germany, FR (under reservation of article 2; 1909), Guatemala (1911), Haiti (1910), Japan (1911), Liberia (1914), Luxembourg (1912), Mexico (1909), Netherlands (1909), Nicaragua (1909), Norway (1910), Panama (1911), Romania (1912), South Africa (1978), Switzerland (1910), Thailand (under reservation of article 1.1; 1910), United Kingdom (1909), USA (1909).

References

Friedman, L. (ed.). 1972. *Law of war: a documentary history.* New York: Random House, 1764 pp.
Roberts, A. & Guelff, R. (eds). 1982. *Documents on the laws of war.* Oxford: Clarendon Press, 498 pp.

Appendix 6. World War II peace or similar treaties relating to sea mines

As a result of the World War II peace or similar treaties imposed upon them, a number of states "... shall not possess, construct or experiment with any ... sea mines ... of non-contact types actuated by influence mechanisms ...".[1] Such a prohibition applies to the following six states:

1. *Austria* by virtue of article 13 of its 1955 Treaty of Independence with the following four states: France, the United Kingdom, the USA and the USSR (Humphrey, 1956, pp. 525–528).

2. *Bulgaria* by virtue of article 13 of its 1947 Treaty of Peace with the following 10 states: Australia, Czechoslovakia, Greece, India, New Zealand, South Africa, the United Kingdom, the USA, the USSR and Yugoslavia (Leiss & Dennett, 1954, pp. 251–272).

3. *Finland* by virtue of article 17 of its 1947 Treaty of Peace with the following eight states: Australia, Canada, Czechoslovakia, India, New Zealand, South Africa, the United Kingdom and the USSR (Leiss & Dennett, 1954, pp. 322–341).

4. *Hungary* by virtue of article 15 of its 1947 Treaty of Peace with the following 10 states: Australia, Canada, Czechoslovakia, India, New Zealand, South Africa, the United Kingdom, the USA, the USSR and Yugoslavia (Leiss & Dennett, 1954, pp. 273–297).

5. *Italy* by virtue of article 51 of its 1947 Treaty of Peace with the following 18 (now 15) states: Australia, Belgium, Brazil, Canada, China, Czechoslovakia, Ethiopia, France,[2] Greece, India, the Netherlands, New Zealand, Poland, South Africa, the United Kingdom,[2] the USA,[2] the USSR and Yugoslavia (Leiss & Dennett, 1954, pp. 163–250).

6. *Romania* by virtue of article 14 of its 1947 Treaty of Peace with the following nine states: Australia, Canada, Czechoslovakia, India, New Zealand, South Africa, the United Kingdom, the USA and the USSR (Leiss & Dennett, 1954, pp. 298–321).

[1] In this context a sea mine actuated by an "influence mechanism" is set off by such influences as the target's magnetic field, by the noise the target generates (its acoustic emanation), by the change in water pressure the target produces, or by other remote signal from the target.
[2] France, the United Kingdom and the USA in 1951 each released Italy from this obligation (Dennett & Durant, 1953, p. 564).

References

Dennett, R. & Durant, K. D. (eds). 1953. *Documents on American foreign relations. XIII. January 1 – December 31, 1951.* Princeton, New Jersey: Princeton University Press, 626 pp.

Humphrey, H. H. (ed.). 1956. *Disarmament and security: a collection of documents 1919–55.* Washington: US Senate Committee on Foreign Relations, 1035 pp. + 1 map.

Leiss, A. C. & Dennett, R. (eds). 1954. *European peace treaties after World War II: negotiations and texts of treaties with Italy, Bulgaria, Hungary, Rumania, and Finland.* Boston: World Peace Foundation, 341 pp.

Appendix 7.   Viet Nam–US Protocols of 1973 concerning the removal of the explosive remnants of war

I.   Introduction

The 'Agreement on Ending the War and Restoring Peace in Viet-Nam' was agreed to in Paris among the Provisional Revolutionary Government of the Republic of South Viet Nam, Democratic Republic of Viet Nam, USA and Republic of Viet Nam[1] on 27 January 1973, and entered into force on the same date (USA & Viet Nam, 1973, pp. 169–188). Of the four separately agreed to protocols accompanying this Agreement, two dealt in whole or part with the explosive remnants of war. These are presented in sections II and III below.

II.   *Protocol concerning the removal of mines in territorial waters*

The 'Protocol to the Agreement on Ending the War and Restoring Peace in Viet-Nam concerning the Removal, Permanent Deactivation, or Destruction of Mines in the Territorial Waters, Ports, Harbors, and Waterways of the Democratic Republic of Viet-Nam' was agreed to in Paris between the Democratic Republic of Viet Nam and USA on 27 January 1973, and entered into force on the same date. The text of the Protocol follows (USA & Viet Nam, 1973, pp. 187–188):

---

[1] The Democratic Republic of Viet Nam (North Viet Nam) in 1976 united with the Republic of South Viet Nam (the successor to the Provisional Revolutionary Government of South Viet Nam, which in 1975 had formally replaced the Republic of Viet Nam [the 'Saigon' regime]) to become the Socialist Republic of Viet Nam.

[2] Agreement on Ending the War and Restoring Peace in Viet-Nam of 1973, article 2: A cease-fire shall be observed throughout South Vietnam as of 2400 hours G.M.T., on January 27, 1973.

   At the same hour, the United States will stop all its military activities against the territory of the Democratic Republic of Vietnam by ground, air and naval forces, wherever they may be based, and end the mining of the territorial waters, ports, harbors, and waterways of the Democratic Republic of Vietnam. The United States will remove, permanently deactivate or destroy all the mines in the territorial waters, ports, harbors, and waterways of North Vietnam as soon as this Agreement goes into effect.

   The complete cessation of hostilities mentioned in this Article shall be durable and without limit of time.

The Government of the United States of America,

The Government of the Democratic Republic of Vietnam,

In implementation of the second paragraph of Article 2 of the Agreement on Ending the War and Restoring Peace in Vietnam[2] signed on this date,

Have agreed as follows:

Article 1

The United States shall clear all the mines it has placed in the territorial waters, ports, harbors, and waterways of the Democratic Republic of Vietnam. This mine clearing operation shall be accomplished by rendering the mines harmless through removal, permanent deactivation, or destruction.

Article 2

With a view to ensuring lasting safety for the movement of people and watercraft and the protection of important installations, mines shall, on the request of the Democratic Republic of Vietnam, be removed or destroyed in the indicated areas; and whenever their removal or destruction is impossible, mines shall be permanently deactivated and their emplacement clearly marked.

Article 3

The mine clearing operation shall begin at twenty-four hundred (2400) hours GMT on January 27, 1973. The representatives of the two parties shall consult immediately on relevant factors and agree upon the earliest possible target date for the completion of the work.

Article 4

The mine clearing operation shall be conducted in accordance with priorities and timing agreed upon by the two parties. For this purpose, representatives of the two parties shall meet at an early date to reach agreement on a program and a plan of implementation. To this end:

(*a*) The United States shall provide its plan for mine clearing operations, including maps of the minefields and information concerning the types, numbers and properties of the mines;

(*b*) The Democratic Republic of Vietnam shall provide all available maps and hydrographic charts and indicate the mined places and all other potential hazards to the mine clearing operations that the Democratic Republic of Vietnam is aware of;

(*c*) The two parties shall agree on the timing of implementation of each segment of the plan and provide timely notice to the public at least forty-eight hours in advance of the beginning of mine clearing operations for that segment.

Article 5

The United States shall be responsible for the mine clearance on inland waterways of the Democratic Republic of Vietnam. The Democratic Republic of Vietnam shall, to the full extent of its capabilities, actively participate in the mine clearance with the means of surveying, removal and destruction and technical advice supplied by the United States.

Article 6

With a view to ensuring the safe movement of people and watercraft on waterways and at sea, the United States shall in the mine clearing process supply timely information about the progress of mine clearing in each area, and about the remaining mines to be destroyed. The United States shall issue a communique when the operations have been concluded.

Article 7

In conducting mine clearing operations, the U.S. personnel engaged in these operations shall respect the sovereignty of the Democratic Republic of Vietnam and shall engage in no activities inconsistent with the Agreement on Ending the War and Restoring Peace in Vietnam and this Protocol. The U.S. personnel engaged in the mine clearing operations shall be immune from the jurisdiction of the Democratic Republic of Vietnam for the duration of the mine clearing operations.

The Democratic Republic of Vietnam shall ensure the safety of the U.S. personnel for the duration of their mine clearing activities on the territory of the Democratic Republic of Vietnam, and shall provide this personnel with all possible assistance and the means needed in the Democratic Republic of Vietnam that have been agreed upon by the two parties.

Article 8

This Protocol to the Paris Agreement on ending the War and Restoring Peace in Vietnam shall enter into force upon signature by the Secretary of State of the Government of the United States of America and the Minister for Foreign Affairs of the Government of the Democratic Republic of Vietnam. It shall be strictly implemented by the two parties.

Done in Paris this twenty-seventh day of January, One Thousand Nine Hundred and Seventy-three, in Vietnamese and English. The Vietnamese and English texts are official and equally authentic.

III.   Protocol concerning the cease-fire in South Viet Nam

The 'Protocol to the Agreement on Ending the War and Restoring Peace in Viet-Nam concerning the Cease-fire in South Viet-Nam and the Joint Military Commissions' was agreed to in Paris among the Provisional Revolutionary Government of the Republic of South Viet Nam, Democratic Republic of Viet Nam, USA and Republic of Viet Nam on 27 January 1973, and entered into force on the same date. The text of the Protocol follows (USA & Viet Nam, 1973, pp. 182–187):

The parties participating in the Paris Conference on Vietnam,
   In implementation of the first paragraph of Article 2, Article 3, Article 5, Article 6, Article 16 and Article 17 of the Agreement on Ending the War and Restoring Peace in Vietnam[3] signed on this date which provide for the cease-fire in South Vietnam and

[3] Agreement on Ending the War and Restoring Peace in Viet-Nam of 1973, article 2, 3, 5, 6, 16 or 17 does not provide a specific basis for article 5 of this Protocol.

the establishment of a Four-Party Joint Military Commission and a Two-Party Joint Military Commission,

  Have agreed as follows:

• • •

Article 5

  (*a*) Within fifteen days after the cease-fire comes into effect, each party shall do its utmost to complete the removal or deactivation of all demolition objects, mine-fields, traps, obstacles or other dangerous objects placed previously, so as not to hamper the population's movement and wrk, in the first place on waterways, roads and railroads in South Vietnam. Those mines which cannot be removed or deactivated within that time shall be clearly marked and must be removed or deactivated as soon as possible.

  (*b*) Emplacement of mines is prohibited, except as a defensive measure around the edges of military installations in places where they do not hamper the population's movement and work, and movement on waterways, roads and railroads. Mines and other obstacles already in place at the edges of military installations may remain in place if they are in places where they do not hamper the population's movement and work, and movement on waterways, roads and railroads.

• • •

Reference

USA & Viet Nam. 1973. Agreement on ending the war and restoring peace in Viet-Nam. *Department of State Bulletin*, Washington, **68**: 169–188. Also: South Viet Nam: Giai Phong Publishing House, 150 pp.

Appendix 8.   Explosive remnants of conventional war: a report to UNEP[1]

Arthur H. Westing *et al.*
Stockholm International Peace Research Institute

I. Introduction

1. This study is the result of a high-level expert meeting, convened at Geneva from 25 to 28 July 1983 by the Executive Director of the United Nations Environment Programme (UNEP). Its purpose is to assist the Executive Director with the problem of material remnants of war, pursuant to United Nations General Assembly resolution 37/215 of 20 December 1982. Various sections and subsections of the present study address the questions raised in that resolution. The study begins with a summary and concludes with recommendations.

II. Summary

2. The following major points emerge from the present study:
   (*a*) Explosive remnants of conventional war have both environmental and

[1] This is the report of a high-level group of eight international experts convened by Dr Mostafa K. Tolba, Executive Director of the United Nations Environment Programme on behalf of the United Nations Secretary-General in Geneva on 25–28 July 1983 under the chairpersonship of Dr Arthur H. Westing of SIPRI. The report appeared as a part of United Nations General Assembly, New York, Document No. A/38/383 (19 October 1983), pages 6–28. (The style of the headings and the like has been brought in line with that of the rest of the book and typographical errors have been corrected.)

The group of experts who prepared the report consisted of: Professor Ali A. Abdussalam (University of Gar Younis, Benghazi); Colonel Bengt Anderberg (Swedish Army, Skövde); Mr Jozef Goldblat (SIPRI); Professor Edward Gordon (Union University Law School, Albany, New York); Professor Mohamed Kassas (University of Cairo); Dr Boguslaw A. Molski (Polish Academy of Sciences, Warsaw); Professor Khairi Sgaier (University of Alfateh, Tripoli); and Dr Arthur H. Westing, chairperson (SIPRI). This group of eight experts also benefitted from the presence of five observers: Mr Yusuf J. Ahmad (United Nations Environment Programme, Nairobi); Mr Abdel-Kader Bensmail (United Nations Department for Disarmament Affairs, Geneva); Mr Paolo Bifani (United Nations Environment Programme, Nairobi); Mr Marcel A. Boisard (United Nations Institute for Training and Research, Geneva); and Dr Mostafa K. Tolba (United Nations Environment Programme, Nairobi).

economic implications. Terrestrial and marine ecosystems can be degraded by such remnants and by the dumping or disposal of unwanted munitions. Remnants of war also impose constraints on the utilization of productive land and of resources; and the costs of clearing mines and other unexploded munitions are high;

(*b*) The residuum of unexploded ordnance remains mortally dangerous for many decades following a war. It is evident that countries on whose territories a war has been fought must maintain highly trained munition-disposal units whose hazardous work must continue unabated the year round for decades;

(*c*) The problem of remnants of war includes technological aspects, among them the need to keep up with continuing developments in mines and booby traps, the need to develop in-built devices of self destruction, and the need for improvements in means of detection and clearance;

(*d*) The problem of remnants of war includes legal aspects, among them the need for world-wide adherence to existing pertinent rules and principles of international law, and the progressive development of these rules;

(*e*) International co-operation is required for: (*i*) information (its collection and dissemination where needed); (*ii*) training of personnel (regarding detection and clearance); and (*iii*) technical assistance to developing countries facing problems of remnants of war (especially so that national capabilities can be achieved); and

(*f*) The letter of UNEP relevant to material remnants of war that was sent in April 1983 to all governments elicited 35 responses by the end of July 1983, including one that contained substantive information on problems related to such remnants. The responses revealed divergent views on: (*i*) the issues of responsibility and liability for compensation; (*ii*) the question of the role of the United Nations and, particularly, of UNEP; and (*iii*) the desirability and feasibility of convening a conference under the auspices of the United Nations.

III. Definitions and estimated magnitudes

Definitions

3. Remnants of war refer to a variety of relics, residuals or devices not used or left behind at the cessation of active hostilities. They include: (*i*) nonexploding devices; (*ii*) unexploded land mines, sea mines and booby traps; (*iii*) unexploded munitions; (*iv*) materials such as barbed wire and sharp metal fragments; (*v*) wreckage of tanks, vehicles and other military equipment; and (*vi*) sunken ships and downed aircraft. The present study focuses primarily on unexploded mines and other unexploded munitions; that is, on the potentially explosive remnants of war. Also considered in brief are similar remnants

of other military activities. Not considered here are the remnants and residues of chemical, biological or nuclear war.

4. Potentially explosive remnants of war have a number of origins. To begin with, there are the munitions that malfunctioned at the time they were expended, the so-called duds. Then there are the sea mines, river mines, land mines and booby traps emplaced, but not subsequently triggered or removed, during the war. Other miscellaneous sources include abandoned ammunition dumps and caches, dumpings of unwanted munitions (often at sea, in lakes or in old mines) and abandoned vehicles, sunken ships, or downed aircraft containing explosive devices or substances.

5. The main potentially explosive remnants of war are:

(*a*) *Mines.* These are the explosive devices usually emplaced and often constructed so as to defy premeditated discovery and designed so as to detonate when disturbed. Mines are usually classified as land, river or sea mines:

(*i*) *Land mines.* These fall into two main categories: anti-tank mines and anti-personnel mines. Other specialized categories include anti-railway, illumination and signal mines. Technical developments in the sphere of land mines are taking place rapidly and involve changes that may lead to new categories; and

(*ii*) *Sea mines and river mines.* These fall into three main categories: contact mines, so-called influence mines (acoustic, pressure and magnetic), and—the most modern ones—moving mines;

(*b*) *Booby traps.* These are explosive charges that are exploded when an unsuspecting person disturbs an apparently harmless object or performs a presumably safe act. Anti-personnel or anti-tank mines can be made into booby traps. Moreover, regular mines of any sort can be booby-trapped in order to make their neutralization more difficult and hazardous; and

(*c*) *Duds.* These are high-explosive munitions that did not burst at the time they were fired.

Estimated magnitudes

6. The assessment of the magnitude of the remnants of war is difficult because often there is no exact information on the location of emplaced mines. Sometimes they have been delivered by means of artillery or aircraft, in which case the ability to record their location becomes haphazard, if not impossible. Unavailability of accurate maps, or geographical and meteorological conditions, may also limit minefield records. The information on unexploded dud munitions is extremely vague; normally there are no records of their location, and their magnitude can be estimated only roughly.

7. Concerning land mines, in the various North African campaigns during World War II, for example, the Allied and the Axis forces laid many millions of such devices, mostly anti-tank mines. The estimates vary from some 5

million to as many as 19 million, according to different sources. In Poland, about 15 million land mines and 74 million other pieces of ordnance have been cleared since World War II. In Finland, about 1 million pieces of ordnance have been disposed of so far. The many wars since World War II continue to add to the problem in various parts of the world.

8. Concerning sea mines, it has been reported that during World War II, the USA, for example, laid nearly 31 000 sea mines in the Pacific Ocean against Japan. This is said to have resulted in the destruction of about 1 100 ships, or one ship for every 28 mines emplaced. It has been further reported that, in the Baltic and North Seas, altogether some 100 000 mines were laid during World War II and some 300 ships were thus blown up. During the Second Indochina War, the USA mined Haiphong harbour from the air with 8 000 sea mines.

9. Concerning unexploded dud munitions, the limited authoritative information indicates that during World War II, from 5 to 10 per cent of all US bombs did not explode, those with delayed-action fuses accounting for the majority of those duds. During the Second Indochina War, US artillery shells equipped with the standard point-detonating fuse failed to explode 2.5 per cent of the time when set in the super-quick mode and from 5 to 50 per cent of the time when set in the delay mode. US mortar shells did not detonate 10 to 20 per cent of the time during the dry season, and 13 to 26 per cent of the time during the wet season. US hand grenades were duds 15 to 25 per cent of the time during the dry season and 40 to 50 per cent of the time during the wet season. The overall failure rate of all high-explosive munitions expended by the USA during the Second Indochina War was estimated to be of the order of 10 per cent.

10. If one employs a dud rate of 10 per cent for purposes of rough estimation in the case of Indochina, one can see that of the order of 2 million bombs, 23 million artillery shells and many tens of millions of other high-explosive munitions did not explode as intended. An unknown fraction of these was salvaged for re-use during the war, often being remanufactured into mines or booby traps. A further unknown fraction was, or has become, sufficiently defective so that the devices will never blow up, and a final unknown fraction remains clearly visible and can thus be avoided and destroyed with relative ease. However, many millions of unexploded munitions remain hidden for long periods as potentially lethal or maiming remnants of that war.

11. The area bordering the Suez Canal was the scene of war in 1967 and the Canal was closed by sunken ships and mines. The Suez Canal area was again the scene of battles during the Attrition War and the armed conflict of October 1973. As a result of those events, immense numbers of undetonated high-explosive munitions got into the Canal and its surrounding land areas; the marine approaches were also mined.

12. Wars of liberation in many developing countries, especially in Africa, have also resulted in the generation of explosive remnants of war. For

example, it has been reported that the Zambezi and Luangwa basins have been mined and that the Lower Zambezi National Park (the former International Game Park), which follows the Zambia–Zimbabwe border, has been virtually abandoned, because it was used as a corridor and heavily mined.

13. More recent military conflicts include the Israeli incursion into Lebanon during 1983, which resulted in the generation of many different types of high-explosive remnants. Ongoing wars between Iran and Iraq, within Afghanistan, and in Central America are without doubt also generating large amounts of remnants of war, including mines.

14. The recent Falklands (Malvinas) War has resulted in the presence of many thousands of undetonated high-explosive munitions. These include thousands of mines and numerous booby traps; in addition, thousands of small plastic anti-personnel mines were scattered indiscriminately from helicopters in unrecorded locations.

15. Certain military activities in peace-time also generate large amounts of remnants that have the same characteristics as the remnants of war. In particular, islands and other sites have been used as test and firing-practice areas, for example, the Puerto Rican island of Culebra, the Hawaiian islands of Manana and Kahoolawe, and the small island of Filfla off the coast of Malta.

IV. Economic and environmental problems and loss of life and property

Environmental problems

16. Material remnants of war can affect ecological balances by disturbing the soil, destroying vegetation, killing fauna, and introducing poisonous substances into the environment. In North Africa, for example, gazelles are reported to have disappeared from sites that were mined during World War II. Remnants of war also degrade the aesthetic value of the environment, including beaches.

17. The terrestrial environment can be seriously affected when remnants of war explode. Such exploding munitions degrade the land through topsoil damage, erosion and in other ways. For example, when a buried 250 kilogram bomb explodes it can produce a crater up to 8 metres across and 4 metres deep. Some of the soil is thrown up and falls back, forming a raised rim around the crater, but much of the soil is compacted into its sides. Some soil washes down to the bottom of the crater, but as vegetation grows it fixes the sides and the crater becomes a virtually permanent part of the landscape. In wet periods, craters may fill with water and become breeding habitats for mosquitoes. The rehabilitation of cratered land is expensive. It takes much effort to fill in a crater and many years for such land to become fully productive again. Trees near an explosion are killed or damaged. It can be added here that if the timber is harvested, shell fragments may damage saws.

18. The marine environment also suffers greatly from the underwater explosion of mines or when unwanted munitions are sunk and blown up. The explosion of a typical depth charge can be expected to be lethal to most marine animals within a radius of perhaps 77 metres, and thus within an area of about 2 hectares and a volume of almost 2 million cubic metres. For fish possessing air (swim) bladders, these values would have to be multiplied by 4, 16 and 64 respectively.

19. A problem closely related to that of the material remnants of war is caused by the disposal at sea of obsolete or surplus munitions. Typically, an unwanted ship is loaded with perhaps 1 million kilograms of such munitions, scuttled in deep water, and often set to explode at a depth of perhaps 700 metres. An explosion of this magnitude is lethal to most marine animals within a radius of more than 1 600 metres, representing a lethal area of greater than 800 hectares and a lethal volume of 20×10^9 cubic metres. Again, as with the depth-charge figures presented above, for fish with air bladders, these values must be multiplied by 4, 16 and 64 respectively.

20. When unwanted munitions are disposed of at sea without being exploded, the toxic properties of their chemical constituents become an environmental hazard. TNT, for example, is heavier than water and sparingly soluble. As the TNT slowly dissolves it kills or inhibits the growth of a number of aquatic micro-organisms and is lethal to some fish. Cyclonite, another important military explosive, is a dangerous mammalian nerve poison; indeed, it is also used commercially as a rat killer. Its half-life in seawater is about 630 days.

21. The process of clearing minefields is not only dangerous, risky and costly, but can also cause excessive damage to the environment. This is especially the case when wide-area explosive methods are employed for neutralizing such areas.

22. In order to minimize adverse environmental impacts of the material remnants of war, it is necessary to develop and make available to affected countries suitable methods of clearing mines and other unexploded munitions. It is necessary for existing detection equipment to be refined and for new types to be developed, especially with reference to non-metallic detection. The problem of explosive munition remnants could be mitigated through the design and adoption of more dependable fuses, which would thus result in the creation of a smaller residuum of dangerous duds. Moreover, every type of high-explosive munition should be designed to have a built-in mechanism for becoming harmless in due course.

Economic problems

23. The economic implications of the material remnants of war are no less serious than the environmental ones. They are of several types, including:

(*a*) Those that follow directly from loss of life and maiming among the productive population; a reduction in livestock; and a loss of property;

(*b*) Those that prevent the use of natural resources or other aspects of the environment either because they have been damaged by the remnants of war or because their use is considered to be dangerous or risky;

(*c*) Those that result from a diversion of resources from productive activities to rectifying the damage caused by the remnants of war; and

(*d*) Those that grow out of a disruption of the social fabric (the disruption of families, loss of income, forced migration, and so forth).

24. The natural environment constitutes the basis of social life and economic development. The direct damage caused by the material remnants of war may therefore destroy the base of socio-economic development. Even the mere existence of remnants of war in certain areas deprives countries of the use of components of their environment, that is, of their natural resources. In sum, unexploded remnants of war endanger people, livestock and wildlife; impede the development of an economic infrastructure (roads, power and telephone lines, airports, etc.); make land unsafe to farm or irrigate; and hamper mineral exploitation. Unexploded remnants of war at sea or in rivers interfere with navigation and with fishing and, if washed ashore, imperil those living along the coast. Bomb craters, wrecked buildings and vehicles, and derelict defences are a blot on the landscape and reduce its value for recreation.

25. Areas that have been mined or otherwise contain unexploded remnants of war become unavailable for economic development or other social pursuits. The alternative to the abandonment of such lands, often large areas, is to undertake clearing activities, which include localization, identification, and neutralization of remnants of war. This overall process requires highly specialized experts and equipment. It is a lengthy process and an extremely risky one, as reflected by the casualties registered during clearing operations.

26. One of the major effects of the material remnants of war is the impediment to the use of large areas of land or sea. For example, the most important part of Libya (in terms of population, agriculture, oil exploitation and industrial activities) is its coastal strip. To a large extent, this region has been severely hampered in its social and economic development because of its World War II mines. Indeed, about 27 per cent of the total arable land of that country is reported to be covered by minefields; and a larger area of the total arable land (68 per cent) is suspected of containing mines and other potentially explosive remnants of war. The development of mineral resources has been prevented as well. Specifically, the development of certain deposits of iron, gypsum, oil, natural gas and potassium salts discovered before the war have remained abandoned to this day because of the dangers involved. Altogether, fully 33 per cent of the entire land area of Libya is considered to be dangerous owing to the explosive remnants of World War II.

27. Oil exploration and development activities have been substantially affected by the material remnants of war in a number of countries, for example, in Libya and Egypt. This has been the result of the large extra costs

resulting from the need to dispose of these remnants from potential oil fields and routes of access.

28. Tourist activity is also affected by the material remnants of war. For example, in Zimbabwe the Lower Zambezi National Park, which covers an area of more than 400 000 hectares, has had to be practically abandoned because the park itself as well as the roads leading to it had been heavily mined during the course of Zimbabwe's war of liberation.

29. Sea mines constitute serious and dangerous problems, and the safety of navigation requires the establishment of fully cleared and marked navigational routes (so-called securely trailed fairway systems). Drifting mines represent serious hazards to exploration and construction activities. Fishing implements are often destroyed or damaged by mine anchors and ship-wrecks. Dumped munitions are occasionally brought to the surface in fishing nets and injure fishermen.

30. The neutralization of sea mines involves sophisticated equipment for detection, including specially equipped ships (minesweepers or minehunters) and, more recently, helicopters. Mechanical, acoustic, magnetic and other sensors are employed. Sea mines are often designed so as to minimize their detectability; it is estimated that some 20 minesweeping passes are necessary before a sector can be considered as clear and safe. Depth charges will set off many, but not all kinds of, sea mines. In shallow coastal and harbour areas it is often considered necessary to supplement ship and helicopter sweeping with detailed underwater searches by divers. Thus, sea-mine clearing remains an extraordinarily time consuming, difficult and dangerous task.

31. The localization, identification and neutralization of land mines and other unexploded munitions also require specialized techniques and highly trained personnel and thus remain hazardous and costly operations. Land mines are often specifically designed and usually specifically emplaced so as to make detection impossible; and dud munitions are often randomly concealed just below the surface. For example, in the years since the end of World War II, mine-disposal units in France have been employing a total of about 90 specialists at any one time, organized into 10 teams, who clear high-explosive munitions on a continuous basis; more than 13 000 were cleared in 1978 alone. Since the end of that war, in West Berlin alone more than 7 000 bombs, more than 718 000 artillery shells, and almost 476 000 grenades and other small explosive devices have been neutralized. Finnish disposal units have, since the end of that war, disposed of over 6 000 bombs, 805 000 artillery shells, 66 000 mines and 370 000 miscellaneous high-explosive munitions.

32. The financial costs related to material remnants of war include not only the ones associated with the direct clearing activities; to those must be added the costs of rehabilitating the affected areas. Further necessary additional costs include expenditures for medical care, rehabilitation, retraining, and the procurement of special equipment and tools for the disabled. A further

economic cost is involved in the need to divert resources from productive economic activities.

Loss of life and property

33. Explosive remnants of war, particularly mines, have caused the loss of much human life. A recent UNEP study based on information provided by governments illustrates the extent of this loss during the three decades following World War II (Tolba, 1977). During that period, in Libya alone the explosive remnants had killed about 4 000 people and injured more than 8 000 others, in both categories most of them children. Additionally, some 460 disposal personnel had been killed in that country and 650 injured. During the last five years, 30 to 40 people have been killed each year in that country, and 50 to 80 injured.

34. The explosive remnants of World War II have had a serious impact on the population of the Netherlands. The post-war hazards have stemmed in part from minefields that were inadequately or otherwise improperly cleared shortly after the war. They also include dispersed underground duds. Three decades after the war about 50 professionals are still engaged daily in clearing those remnants, which continue to cause casualties among both the disposal personnel and the civilian population.

35. The residuum of unexploded ordnance remains mortally dangerous for many decades following a war. For example, in North Africa during the four decades since World War II, it has continued to be no rare incident for shepherds to step on old buried mines and to be killed in the explosion that followed. In a tragic incident in Burma in 1976, 21 people were killed and about 300 injured when a World War II shell exploded in a village near Mandalay. As another recent example, the explosion of a war remnant killed five pupils in a school near Strasbourg, France.

36. Vast amounts of munitions were expended in South Viet Nam during the Second Indochina War. These resulted in uncounted millions of dangerous duds, among them a variety of delayed-action anti-personnel weapons. The combatants employed anti-tank mines, anti-personnel mines and booby traps, so that very large numbers of these devices also remain. There are many places in the highlands where it is too dangerous to enter. Moreover, a great number of rural families can recount personal tragedies, whether of death or maiming, caused by previously unexploded munitions.

37. It is evident that countries on whose territory a war has been fought must maintain highly trained munition-disposal units whose dangerous work must continue unabated the year round for decades. The casualty rate among such personnel is estimated to be one killed and two wounded for about every 5 000 mines rendered harmless. To provide a concrete example, in one recent operation, Egyptian disposal units experienced one fatality for every 7 000 land mines removed.

38.   Important among the other resources affected by the remnants of war
are livestock. In Libya, for example, during and since World War II, more
than 125 000 domestic animals (camels, sheep, goats, cattle) have been killed,
of which about 60 per cent were camels. Associated water points were also
affected and their rehabilitation entailed additional costs. Remnants of war
also continue to constrain the utilization of range and other arable lands.

V.   Demands of affected countries and extent to which responsible states are willing to compensate and assist those countries

Demands of affected countries

39.   The long-term problem of material remnants of war, particularly of
mines and other unexploded ordnance, is a very grave one. Many of the
countries affected were under foreign domination at the time of emplacement
of the devices. The technological advances relating to munitions make for an
increasingly difficult problem for these, and indeed all, countries to cope
with.

40.   The demands of affected countries for fair compensation must be
seen in the light of the suffering and devastation, including loss of lives and
property, experienced by them. Moreover, future demands will be affected
by technological advances that make mine detection increasingly difficult
and that give mines increasing longevity; by the nature of new weapons, such
as cluster bombs and scatterable mines; and, above all, by the growing
emphasis on large-scale area neutralization or denial practised in modern
warfare. The problem is further compounded by the reticence of the
countries that employ sophisticated land and sea mines to disclose relevant
technical details, and by the increasing difficulty in keeping track of where
mines have been placed.

41.   Developments in science and technology are unbalanced by a notable
increase in the efficiency of weapons of destruction in contrast to a slower
development in the technological aspects of avoiding or counteracting such
destructive potential. In other words, mine technology has advanced
considerably beyond the technology of mine detection and neutralization.

42.   In these circumstances, it is possible to conclude that the hardships
that certain countries are currently undergoing because of remnants of war
require, in particular:

(a)   Provision of adequate maps that show the location of the minefields;
(b)   Furnishing of necessary information about the types of mines laid in
      different land locations and territorial waters;
(c)   Provision of techniques and expertise to help locate and neutralize the
      mines; and
(d)   Payment of compensation to the countries affected for the loss of lives
      and property, and for other damages, caused by the remnants of war.

Extent to which responsible states are willing to compensate and assist affected countries

43. The question of compensation has been indicated by some governments as an issue that may delay or jeopardize the achievement of any practical solution or agreement regarding the problem of material remnants of war. Therefore, they suggest that in the discussion and search for a solution to this problem, the question of compensation should be left aside.

44. It is the expressed opinion of three governments that the solution of the question of compensation is a subject for bilateral agreements. Three governments do not consider UNEP to be the appropriate forum for the discussion of this matter.

45. The problem of identification of responsible states is made extremely difficult owing, among other reasons, to a changing of the geopolitical situation following conflicts. From the replies received from governments, it appears that no government is prepared to assume responsibility for remnants of war; therefore, the willingness to compensate has not been mentioned. Nevertheless, the important issue of responsibility for damage and compensation should not be minimized or neglected. All co-operative arrangements for clearance of remnants of war are welcome, especially those based upon a fair measure of compensation.

46. Internal conflicts and wars of liberation create particular problems of responsibility regarding the remnants of war.

47. Many developed countries have carried out disposal operations since World War II and there are, thus, a great number of experienced personnel and much special equipment available in those countries. For example, in 1973, the USA agreed to sweep Haiphong harbour [see appendix 7], a five-month job by a large naval task force that was not as difficult as it might have been, inasmuch as the mines had been laid by the USA with subsequent relocation in mind. The mines sown in inland waters did not have to be sought out because they had been set to destroy themselves or become inert after a time.

48. In the case of the recent Falklands (Malvinas) War, British disposal units have already cleared thousands of the explosive remnants of this brief conflict, but the work is expected to continue for at least another year in the farming regions and for years beyond that in the peat bogs (which need to be exploited for fuel) and other rural regions. At least one of the disposal personnel has been killed so far and several have lost limbs.

49. In recent years there have also been cases of international collaboration, among them the combined and very successful operation in clearing the Suez Canal and the ongoing United Nations activity in Lebanon. Those cases have shown the multifaceted nature of the effort needed and the importance of international co-operation.

50. The clearing of the Suez Canal required a huge, sophisticated and dangerous series of aerial, surface and subsurface operations by Egypt, the

USA, the United Kingdom, France and the USSR over a period of more than a year to render finally the Canal and its approaches sufficiently safe to be dredged and reopened. The United Kingdom contingent, during a single five-month period, found resting on one stretch of canal bed, and neutralized, 516 anti-personnel mines, 125 anti-tank mines, 16 bombs, 9 cluster bombs, 508 bomblets, 234 artillery shells, 141 anti-tank rockets, 190 grenades and many hundreds of miscellaneous additional items of explosive ordnance. All told, some 8 500 diverse items of explosive ordnance were found in the Canal and disposed of. Moreover, the Egyptian contingent cleared nearly 700 000 land mines from the terrain adjacent to the Canal. Many of these mines were non-metallic and had to be located manually since the best available electronic detectors had not been adequate for the job.

51.   The aftermath of the recent Israeli incursion into Lebanon includes the location and disposal problems associated with large numbers of many different types of high-explosive remnants. A joint US–Lebanese munition-disposal unit (one of several) in a six-week period unearthed 250 different kinds of explosive ordnance, including more than a dozen bombs, some 200 bomblets, and hundreds of mines and grenades. Forty-five bomblets were disposed of in the yard of an orphanage after an explosion that killed four children and wounded five others. As noted earlier, children are often the victims of such tragedies. Several US and French disposal personnel have also been killed so far in the line of duty.

VI. Legal aspects

52.   The legal aspects of material remnants of war are complex and subject to widely differing interpretations.

53.   State responsibility with respect to material remnants of war may result when weapons are involved, the use of which is prohibited or restricted under international law, and when, following the termination of hostilities, remnants of war present a continuing peril that calls for international co-operation in their removal.

54.   International law imposes few outright prohibitions on the use of specific weapons of relevance to the present question. Hague Convention VIII of 1907 [see appendix 5] limits the laying of automatic contact sea mines and requires combatants to take every possible precaution for the security of peaceful shipping. The combatants are also required to do their utmost to render the mines harmless within a limited time and, should these cease to be under surveillance, notification is to be given to ship owners and governments of the danger zones as soon as military exigencies permit. Non-combatant states laying mines off their own coasts must observe the same rules. Upon the conclusion of hostilities, each party is obliged to remove the mines that it has placed in its own waters and to make known to the other party the position of mines it has placed off that party's coasts.

55. The 1971 Treaty on the Prohibition of the Emplacement of Nuclear Weapons and other Weapons of Mass Destruction on the Seabed and the Ocean Floor and in the Subsoil thereof prohibits the stationing of any nuclear weapon or any other type of weapon of mass destruction on the seabed and the ocean floor and in the subsoil thereof beyond the outer limit of a specified seabed zone. Implicit in this prohibition is the proscription of nuclear mines, as well as mines containing chemical and biological warfare agents, anchored to or installed on the seabed.

56. Protocol II of the Inhumane Weapons Convention of 1981 [see appendix 3] imposes restrictions on the use of land mines, booby traps and certain analogous weapons of relevance in the present context. It requires that all feasible precautions be taken to protect civilians from the effects of these weapons. It also requires the recording of the location of minefields, mines, and booby traps, and the provision of this and related information to other combatants following the termination of hostilities and the removal of forces. Protocol II also contains rules designed specifically to protect United Nations peace keeping, observation, or similar forces or missions from the perils posed by unexploded remnants of war.

57. Pertinent principles of general international law prohibit the use of weapons indiscriminately without regard to the safety of non-combatants, as well as the resort to means and methods of warfare that are of a nature to cause superfluous injury or unnecessary suffering or that are disproportionate to the military objective. Especially relevant are the principles regarding the protection of the civilian population, including those prohibiting action expected to cause incidental loss of life or injuries among civilians, damage to civilian objects, or denial of objects that are indispensable for the survival of the population.

58. It is the opinion of some that international law, in common with trends perceptible in domestic law, has begun to impose liability for failure to control adequately certain conduct that may not be regarded as wrongful *per se*, but which involves substantial and foreseeable risk to others or to fundamental values of the community. With a view to allocating fairly the burden of loss, several recent treaties and international decisions provide, in specific contexts, that the way in which states use or manage their physical environment, either within their own territory or in areas within their effective control, may give rise to international liability for injurious consequences, notwithstanding that the use itself is not prohibited under international law and may even be in furtherance of desirable ends. In the opinion of others, direct analogies should not be drawn between damages caused in time of peace, which are clearly subject to compensation, and those that result from armed conflict, which may or may not be.

59. Under the principles of the Charter of the United Nations, particularly as reflected in the 'Declaration of Principles of International Law concerning Friendly Relations and Co-operation among States in accordance

with the Charter of the United Nations', states are obligated to co-operate with one another irrespective of the differences in their political, economic and social systems, in the various spheres of international relations (UNGA, 1970). This may be said to imply that humanitarian imperatives should be accorded precedence over military and political considerations, especially when remnants of war pose a threat to the health or survival of the civilian population. Accordingly, in the interest of developing countries, and of deterring impending environmental harm and minimizing existing environmental damage, the co-operation of developed states that are in a position to assist in the removal of explosive remnants of war would seem to be called for. Indeed, Protocol II of the Inhumane Weapons Convention of 1981 (see para. 56) envisages the possibility of agreements for joint operations to remove or render ineffective mines placed during armed conflict. However, the present nature of the obligation of states to co-operate is not a matter on which a consensus exists.

60. Individual disputes tend to involve particular legal considerations and thus resist statements of a general nature. Moreover, it should be noted that legal claims, otherwise valid, may be regarded by certain states as having lapsed, or become unenforceable, owing to the passage of time.

61. The Geneva Convention IV of 1949 Relative to the Protection of Civilian Persons in Time of War implicitly prohibits compelling protected persons to participate in mine-removal operations, while the Geneva Convention III of 1949 [see appendix 4] explicitly prohibits the involuntary use of prisoners of war in such operations.

62. The Hague Convention IV of 1907 Respecting the Laws and Customs of War on Land established the principle—subsequently reiterated in the 1977 Protocol I Additional to the Geneva Conventions of 1949, and Relating to the Protection of Victims of International Armed Conflicts—that a party to a conflict that has violated the Conventions or the Protocol should be liable to pay compensation. However, peace treaties and armistice and cease-fire agreements, which might be expected to provide evidence of state practice with respect to reparations for injuries resulting from breaches of rules and principles governing the use of means and methods of war, in fact defy a search for regularity and consistent underlying logic. The justification for reparations is not always given nor are reparations arising from the breach of legal obligations treated separately from reparations merely exacted from the vanquished state. Moreover, no peace agreement imposes a duty to make reparation on the victorious states or acknowledges the victors' own responsibilities under international law. Some agreements reflect a willingness on the part of former combatants to assist one another in locating, identifying and disarming explosive remnants of war, including land mines, but these agreements do not appear to possess the degree of uniformity, frequency or recognition of legal compulsion necessary to warrant the conclusion that such practice constitutes or reflects a rule of customary international law.

63.   It has been advocated that the following principles be taken into account in any legal consideration of the material remnants of war:

(*a*) International co-operation and good neighbourliness;

(*b*) Equity;

(*c*) Refraining in international relations from the threat or use of force against the territorial integrity or political independence of any state, or from any other inconsistency with the purposes of the United Nations (UN Charter, article 2.4);

(*d*) Permanent sovereignty over natural resources;

(*e*) State succession and the law of decolonization; and

(*f*) Objective or strict liability.

VII.   International co-operation required to solve the problem, including the role of the United Nations

Required international co-operation

64.   International co-operation in the field of material remnants of war should be considered under two main headings: (*a*) how to tackle the existing problems in developing countries; and (*b*) how to establish preventive measures for the future.

65.   International co-operation for the clearing of existing remnants of war and the repair and recovery of the damage inflicted can be initiated in terms of different types of action, among them: the providing of information; legal assistance; technical and economic assistance; research and pilot studies; and joint clearance and rehabilitation operations.

66.   One of the main problems involved in the removal of remnants of war involves their detection and identification. Information is required on the exact location of emplaced mines, the numbers of such mines and their types. When available, such information can be compiled into registers, including maps.

67.   A second aspect concerning information refers to the methods and technology for dealing with different types of war remnant, specifying the characteristics of each technique and for which type of mine or dud and under what conditions it can be used, as well as its cost.

68.   In its Decision 6/15 of 15 May 1978, the Governing Council of UNEP requested all governments in possession of the appropriate technology for dealing with environmental hazards caused by remnants of wars to register relevant sources with the International Referral System; and requested the Executive Director of UNEP to continue to gather, through the System, sources of information on this subject matter. Such a clearing house and a repository should be established and the gathering of information actively pursued.

69. Technical and economic co-operation and assistance can be on a bilateral or a multilateral basis. Such technical assistance has two levels: the first, the preparation of specific programmes for the elimination of remnants of war; and the second, the clearance process itself. The first level of technical assistance can be considered as an integral part of development projects for the rehabilitation of areas affected by remnants of war.

70. International co-operation in respect of the clearing process itself can take the form of joint operations, with two or several parties co-operating. The successful joint operation for the clearance of the Suez Canal noted earlier provides an interesting model.

71. It is necessary to carry out further studies in order to assess the environmental effects and the economic implications of different types of mines and explosive devices in relation to specific ecosystems. These studies should be carried out within the framework of the impact of military activities on environment and development.

72. The problems posed by explosive remnants of war could be mitigated by more widespread adoption and adherence to the relevant treaties, especially the Inhumane Weapon Convention of 1981 and the Protocols thereto [see appendix 3]. International co-operation should aim at the prevention and avoidance of environmental hazards and of the negative impact on the development process.

Role of the United Nations

73. The gravity of the situation described in this report makes it important, indeed necessary, for the international community to contemplate adequate ameliorative action. The asymmetrical technical capacities of different countries—the affected developing countries lacking in technical expertise for the most part, and the developed countries having an abundance of human skills and technology—make it advisable that such action be taken under the aegis of the United Nations system.

74. A number of actions have become increasingly urgent and should be taken in hand without further delay. One of them is the collection, classification and categorization of information on remnants of war: their location, their magnitude, their nature and their destructive capacities. Collection of information on the nature of new munitions that will lead to remnants of war is also necessary. A register of information could be established by a United Nations body for this purpose with contributions from all countries. Similarly, it has become a major priority to establish a data-base of information on techniques and expertise available for the removal of the remnants of war.

75. A promising role for the United Nations system is to establish modalities for the channelling of technical and financial assistance towards the removal of the remnants of war.

76. The legal issues involved which are fundamental to the proper resolution of the problem need to be seriously considered by the appropriate bodies of the United Nations system.

Possibility of convening a conference under United Nations auspices

77. In its resolution 37/215 of 20 December 1982, the United Nations General Assembly requested the Secretary-General, in co-operation with the Executive Director of UNEP, to prepare a study on the problem of the remnants of war which would include an analysis of the role of the United Nations in this regard, including the possibility of convening a conference. Such a meeting would serve an important purpose as a clearing house of ideas and policies and would undoubtedly help to lay the foundation for future co-operative action in terms of actual mine-clearing operations.

78. If a conference under United Nations auspices were convened, it could be organized in one or more of the following three formats:

(*a*) A United Nations conference covering the full spectrum of issues. Such a conference, however, could give rise to a number of potential difficulties. Issues related to the material remnants of war have been extensively examined in the past and member states have been given an opportunity to express their views. On the basis of past expressions, it appears unlikely at this time that a consensus would emerge on the issues over which government opinions were clearly divergent;

(*b*) A United Nations conference covering a limited spectrum of issues. Such a conference could deal with transfer of relevant technology, assistance for removal, and so forth; and

(*c*) A United Nations-sponsored meeting of a group of government-nominated experts (around 20 with proper geographical distribution) to cover either the totality of issues or a limited agenda (for example, technical issues on methods of clearing mines) in order to make specific recommendations for action by the United Nations General Assembly.

VIII. Recommendations

79. The United Nations General Assembly may wish to make a strong appeal to all states for urgent remedial action on the problem of material remnants of war, emphasizing in particular the aspects that are described below.

Legal aspects

80. The United Nations General Assembly may wish to appeal to all states to ratify or accede to the Inhumane Weapons Convention of 1981 and the relevant Protocols thereto [see appendix 3] and to request the Secretary-General to report periodically on the status of the implementation of that Convention. On the first appropriate occasion, consideration should be given to expanding the scope of this Convention, for example, by bringing up to date existing rules and principles regarding sea mines and developing new rules for dealing with the question of dumping stocks of munitions in the ocean.

81. The legal issues presented by material remnants of war are extremely complex. Widely differing views exist on matters of responsibility and the possibility of fair compensation. Thus, in quest of securing an authoritative pronouncement of pertinent legal rules and principles regarding material remnants of war, the United Nations General Assembly may wish to encourage the International Law Commission to consider relevant legal issues and to request the International Court of Justice to render an advisory opinion.

82. States that are in disagreement concerning legal obligations to assist in the removal of material remnants of war may be encouraged by the United Nations General Assembly to resolve their disagreements in accordance with the principles contained in the Charter of the United Nations. Such states may also consider resolving their disagreements through third-party arbitration or adjudication, especially through the good offices of the Secretary-General of the United Nations and/or the Executive Director of UNEP.

Informational aspects

83. Technical and historical information is required about the areas and objects to be cleared. The collection, classification, categorization and dissemination of information on the extant remnants of war deserve special attention. In spite of international efforts undertaken during the past several years, the availability of pertinent information of a factual and technical nature has remained sporadic in character and scope. The United Nations General Assembly may wish to urge all states to co-operate more effectively and fully in the referral arrangements initiated by UNEP and, going beyond referral, by supplying adequate maps that show the location of minefields and furnishing information about types of mines laid.

84. Similarly, there is an urgent need for a data-base on technologies currently available for mine clearance. It is a matter of concern that, in an area of international co-operation where human considerations should be paramount and respect for human life should take precedence, there still remains so much reluctance to exchange data and technical details. Thus, the United Nations General Assembly may wish to recommend strongly to all states that they co-operate in the creation and upkeep of an appropriate and adequate data-base.

Technical assistance

85. The work of clearing material remnants of war is hazardous, time-consuming, costly and could be damaging to the environment. It is now not technically feasible to clear large affected areas simply and quickly. Even if very substantial resources are allocated, one can never guarantee that an area is completely cleared.

86. In view of the asymmetrical capabilities of member states, technical assistance and co-operation under United Nations auspices, and/or in other fashion, is required urgently in this area. Past experience illustrates the value of international co-operation in clearance operations. Also needed is the establishment of training programmes for personnel from developing countries. The United Nations should also endeavour to assist developing countries either directly or through helping in channelling bilateral assistance in upgrading their technical equipment needed for detecting and clearing remnants of war.

87. Because of the rapid technical development of weapon systems, the problems of material remnants of war may be accentuated in the future. With ever more widespread use of wide-area weapons, such as those that contain many sub-warheads, the danger of duds has dramatically increased. Also, the emplacement of large numbers of advanced sophisticated mines can have serious effects. The economic and environmental consequences of material remnants of war indicate that it is much more economical to clear such remnants promptly than to allow them to remain in place. It is therefore urged that preparations be made to increase the possibilities of quickly clearing the explosive remnants of future conflicts.

88. Suitable technology for clearing explosive remnants of war must be developed and made readily available to affected countries.

89. High-explosive munitions should be designed to have built-in mechanisms that render the munitions harmless in due course.

Institutional arrangements

90. Clearance of the material remnants of war that constitute a threat to the environment should be carried out, as appropriate, through international co-operation, preferably under the aegis of the United Nations. This calls for action by a United Nations body—presumably one already in existence—that would be able to collect information and data relevant to eliminating the perils from the material remnants of war; and on material, technological and human resources that are available. Such a body should, at the request of and in co-operation with the countries affected, make proposals on how clearance operations can take place. To meet such a requirement, this body should be able to follow the technical developments in the relevant fields.

91. In this respect, consideration should be given to voluntary earmarking by governments of pledges in human and other resources through the United Nations body in question so that they may be used upon request in a timely manner.

92. Other mechanisms for channelling technical and financial assistance to operational activities in the clearing of mines and the removal of other explosive remnants of war should also be considered, especially those including voluntary contributions.

United Nations conference/meeting

93. The considerations raised above are such as to deserve a structured response from the United Nations system. The United Nations General Assembly may wish, in this connection, to organize an international conference to discuss this complex and multifaceted problem. Such an international conference could be organized in different ways. The following options were elaborated in a prior section (see para. 78):

(*a*) A United Nations conference covering the full spectrum of issues;

(*b*) A United Nations conference covering a limited spectrum of issues; and

(*c*) A United Nations-sponsored meeting of a group of government-nominated experts (say 20 in number) to cover the full spectrum of issues or a restricted agenda of issues which could then come to the United Nations General Assembly with specific recommendations for action.

Other recommendations

94. Because of the present imbalance between the development in science and technology regarding the efficiency of weapons of destruction compared to those aspects of avoiding or counteracting such destructive potential, it is essential that research into mine detection and neutralization technology be expanded far beyond its present state.

95. The important issues of responsibility for damage and compensation should not be minimized or neglected. Fair compensation must be considered in the light of damage and suffering entailed by remnants of war.

References

Tolba, M. K. 1977. *Implementation of General Assembly resolution 3435 (XXX): study of the problem of the material remnants of wars, particularly mines, and their effect on the environment.* Nairobi: UN Environment Programme Document No. UNEP/GC/103 (19 Apr 1977), 8 pp. + UNEP/GC/103/Corr.1 (6 May 1977), 1 p. Also: New York: UN General Assembly Document No. A/32/137 (27 Jul 1977), 1 + 8 pp.

UNGA (United Nations General Assembly). 1970. *Declaration of principles of international law concerning friendly relations and co-operation among states in accordance with the charter of the United Nations.* New York: United Nations General Assembly Resolution No. 2625 (XXV) (24 Oct 1970), 12 pp. Reprinted in: *UN Yearbook*, New York, **24**: 788–792.

Index